Engineering Analysis with MAPLE®/MATHEMATICA®

Engineering Analysis with MAPLE®/MATHEMATICA®

Abraham I. Beltzer
Holon Center for Technological Education
affiliated with Tel-Aviv University
Holon
Israel

ACADEMIC PRESS
Harcourt Brace & Company, Publishers
London   San Diego   New York
Boston   Sydney   Tokyo   Toronto

ACADEMIC PRESS LIMITED
24–28 Oval Road
LONDON NW1 7DX

U.S. Edition published by
ACADEMIC PRESS INC.
San Diego, CA 92101

Copyright © 1995 by ACADEMIC PRESS LIMITED

MATHEMATICA is a registered trademark of Wolfram Research Inc.
MAPLE is a registered trademark of Waterloo Maple Software.

This book is printed on acid-free paper

All rights reserved
No part of this book may be reproduced or transmitted in any form or by any means, electronic or mechanical including photocopying, recording, or any information storage and retrieval system without permission in writing from the publishers

A catalogue record for this book is available from the British Library

ISBN   0-12-085570-4

Typeset by T & A Typesetting Services, Rochdale, Lancashire
Printed and bound by Hartnolls Ltd., Bodmin, Cornwall

Contents

Preface x

M. Symbolic Manipulation Codes – Introduction to
   MAPLE/MATHEMATICA. 1
   - M1. MAPLE – first sessions 1
   - M2. Repetitions, conditionals, expansions and integrations 6
   - M3. Arrays and matrices 9
   - M4. Solving equations. 11
   - M5. Hints 13
   - M6. MATHEMATICA – first sessions 15
   - M7. Tables and matrices. 19
   - M8. Solving equations. 21
   - M9. Hints 22

D. Direct Methods 25
   - D1. The functional and its stationarity 26
   - D2. Hamilton's principle 28
   - D3. Minimum of potential energy 31
   - D4. The Bubnov–Galerkin method 32
   - D5. Beam on elastic foundation 35
   - D6. The Rayleigh–Ritz method 36
   - D7. Master program. 38
   - D8. Applications 40
   - D9. Considerations of accuracy 41
   - D10. Mathematical considerations 43
   - D11. Operators 44
   - D12. Energy convergence and minimum theorem 46
   - D13. More on trial functions 48
   - D14. Methods of weighted residual 49
   - D15. The iterative Kantorovich method 50

vi CONTENTS

    D16. Heat transfer in a plate 52
    D17. Constraints. 54
    D18. Computer-generated Euler–Lagrange equations 56
    D19. Two degrees of freedom. 58
    D20. Continuous systems 59
    D21. Automatic derivation of governing equations 61
    D22. Nature of extremum. 63
    D23. Transversality conditions. 64
    D24. Generalizations 66
    D25. Gauss' principle. 67
    D26. Minimum pressure drag. 68
    D27. Approximating boundary conditions. 71
    D28. Use of logical expressions 74
    D29. Concluding remarks 76

F. Finite Element Method 77
    F1. Element stiffness matrix. 77
    F2. Energy analysis 82
    F3. Truss element 85
    F4. Physical meaning of matrices 90
    F5. Eigenvalues of stiffness matrix 92
    F6. Reference systems. 95
    F7. Generalizations 98
    F8. Assembling. 99
    F9. Assembling via equilibrium equations 101
    F10. Assembling via connectivity 105
    F11. Applications to a truss. 108
    F12. Inhomogeneous beam. 114
    F13. Applications to a beam 116
    F14. Automatic generation of an assembly matrix. 117
    F15. Optimization 121
    F16. Imposition of constraints. 124
    F17. Free vibrations 128
    F18. Plate element. 132
    F19. Programming for plate element 134
    F20. Applications. 137
    F21. Two-dimensional elasticity, 1 140
    F22. Two-dimensional elasticity, 2 145
    F23. Heat conduction and related problems 148
    F24. The Bubnov–Galerkin formulation 150
    F25. Heat transfer in a fin 152
    F26. Numerical example. 154
    F27. Time-history analysis 158

F28.	Numerical example	160
F29.	Natural reference systems	163
F30.	Serendipity coordinates	168
F31.	Concept of isoparametric elements	170
F32.	Three-node isoparametric bar element	173
F33.	Two-dimensional isoparametric element	175
F34.	Programming for an isoparametric element	177
F35.	Special elements	179
F36.	Constraints and Lagrange multipliers	182
F37.	Constraints and penalty parameters	184
F38.	Accuracy, convergence and related subjects	186
F39.	More on element deficiency	188
F40.	Symbolic database	189
F41.	Concluding remarks	192

S. **Finite Difference Methods** 193

S1.	FD–operators and their accuracy	193
S2.	More about accuracy	195
S3.	Sample problem	197
S4.	Runge–Kutta methods, 1	199
S5.	Runge–Kutta methods, 2	201
S6.	Transient vibrations	203
S7.	Multi-step schemes	205
S8.	Automatic generation of multi-step schemes	207
S9.	Start-up for multi-step schemes	209
S10.	Iterative predictor–corrector methods	211
S11.	Non-linear vibrations	212
S12.	Newmark's method	215
S13.	Finite difference equations	217
S14.	Stability	219
S15.	Solving non-linear equations	221
S16.	Boundary-value problems	224
S17.	Discretization of PDEs	225
S18.	Explicit scheme	226
S19.	Implicit scheme	228
S20.	Stability	229
S21.	Consistency	230
S22.	Other FD-schemes	231
S23.	Treatment of boundary conditions	232
S24.	Heat conduction in a slab (explicit scheme)	234
S25.	Heat conduction in a slab (implicit scheme)	235
S26.	Heat conduction in a solidifying alloy, 1	237
S27.	Heat conduction in a solidifying alloy, 2	239

viii CONTENTS

  S28. Treatment of irregular boundaries........................ 241
  S29. Torsion of a shaft, 1................................... 241
  S30. Torsion of a shaft, 2................................... 244
  S31. Concluding remarks.................................... 246

W. Workshop... 247
  W1. Analysis of plates, 1................................... 247
  W2. Analysis of plates, 2................................... 250
  W3. Analysis of a shock-absorber, 1.......................... 252
  W4. Analysis of a shock-absorber, 2.......................... 254
  W5. Flow through a duct 256
  W6. Free vibrations by the Rayleigh–Ritz method 258
  W7. Applications, 1 259
  W8. Applications, 2 260
  W9. Free vibrations by the Bubnov–Galerkin method 262
  W10. Non-linear free vibrations.............................. 264
  W11. Applications of Gauss' principle 266
  W12. Heat transfer to fluid................................. 268

Problems.. 271
Appendix A: Manipulations with Matrices......................... 275
Appendix B: Section-file Correspondence 277
Further reading.. 279
Index... 281

PREFACE

The variational, finite element and finite difference methods constitute the very core of engineering analysis, but the associated computations are tedious at best, and they often obscure both the ideas and techniques of the approach.

A resort to symbolic computation codes makes it possible to reverse the situation. The ease with which they provide analytical results allow a student or researcher to focus on the ideas rather than on calculation difficulties. The very process of programming now encourages the appreciation of qualitative aspects of investigations. Saving time and effort, these codes enable undergraduates to deal with advanced subjects. Their semantic features often allow for generalizations of standard techniques, which would be impossible to achieve without the computer's help. There is a habitual aspect too. These days, it is more convenient for a student (researcher) to work with a keyboard than with a pencil. Surprisingly enough, the same digital computers, which in the past brought about an overemphasis on the use of 'discrete' numerical methods, now help to restore the equilibrium and enhance our potential in performing analytical transformations. This promises to be particularly fruitful.

Nevertheless, symbolic manipulation codes may not provide a solution in its most concise and elegant form, nor do they ensure its correctness can be taken for granted. This author feels that some of his peers may successfully compete with the codes from this viewpoint.

It has been recently realized that the overemphasis on mathematical techniques may have a damaging effect as far as engineering is concerned. Indeed, the primary task of engineering schools is to provide the technological, not the computational knowledge. Wide use of symbolic manipulation codes may be extremely useful from this point of view, as they allow for a larger 'portion' of mathematics to be taken over by computers. It would be rational to use these codes at the final stage of undergraduate or during graduate studies, thereby allowing the students to acquire the basic mathematical techniques in their first years.

x   PREFACE

Though a symbolic manipulation code (computerized algebra) is an organic part of this book, it is a tool, not a subject. A collection of the allied computer programs is given in both MAPLE and MATHEMATICA. This provides a possibility for the user to turn to either or both of them and draw conclusions of their own about their suitability for his or her purposes. A considerable analytical database is given, which may be further extended and modified by readers for their own purpose and taste.

The text consists of five chapters denoted (for the sake of convenience) by the letters M (Maple/Mathematica), D (Direct), F (Finite), S (Schemes) and W (Workshop), respectively. The presentation is based on MAPLE.

The first chapter (M) presents the basic information on the symbolic codes. Chapter D introduces the basic concepts of the variational approach and direct techniques, such as the Bubnov–Galerkin and Rayleigh–Ritz methods. Some rarely discussed approaches like Gauss' principle and the iterative Kantorovich method are also given. It is shown that these classical techniques may be reformulated in a somewhat generalized way, if performed with the help of a symbolic code. Chapter F presents the finite element method. This technique, which has been traditionally thought of as a purely numerical one, may provide analytical results, when performed with the help of a symbolic manipulation code. Automatic derivations of relevant matrices are given, among others. Chapter S deals with the finite difference approach, considering both ordinary and partial differential equations. Once again, a combination of analytical, numerical and graphical capabilities provided by computer algebra systems comes heavily into play. Eventually, a set of practical solved problems is given in Chapter W.

The allied programs (*vfem*-files) are written for MAPLE and MATHEMATICA under DOS, Windows, VMS or UNIX. The *vfem*-files for MAPLE have the extension .ms and those for MATHEMATICA the extension .ma. Modifying these files with the help of an editor, the reader may experiment with a variety of problems. These files should be placed under proper directories. For example, to run under MAPLE V Release 2 a proper directory is `maplev2` and to run under MATHEMATICA version 2.2 a directory is `wnmath22`. The section-file correspondence is given in Appendix B.

Further, some programs allow for an automated derivation of analytic expressions, which may be used by the reader for a convenient creation of his/her own database. In a sense, the mutually integrated book and programs may be viewed as a computerized system which enables the user to study, teach and experiment in the broad field of variational, finite element and finite difference methods.

Seven years ago, when I wrote my previous book on variational and finite element methods performed with the help of computer algebra, such codes as MACSYMA, MAPLE and MATHEMATICA were not widely available for PCs. Since then these symbolic manipulators have undergone a spectacular improvement and now run on practically any modern home computer. (At present AXIOM is available only for workstations.) There is little doubt that symbolic computation and electronic interactive books will become common in the near future, as the convenience they offer is substantial and the means available usually affect our way of thinking.

Notation and software comments

There are two systems of notation, in order to allow for a use of earlier versions of MAPLE which do not support Greek letters. One is a traditional system employing English and Greek letters, which is customary for scientific and technical literature. The second system (for the embedded symbolic computation) uses English letters only. The same quantity may be designated, say, as Y_1 in the first case and Y1 in the second. In reproducing MAPLE results, I often used the so-called lprint (one-dimensional) mode, which allows for a more compact printout.

The disk contains the batch files for MAPLE V Release 2 under directory r2, for MAPLE V Release 3 under directory r3, and for MATHEMATICA 2.2 under directory mat. For a specific section-file correspondence, consult Appendix B.

Note that there are differences between the performance of MAPLE V Release 2 and Release 3, in particular with procedures, integration and notation. These do not cause any substantial difficulties in the context of this book.

The batch files for MATHEMATICA contain extensive comments to facilitate their association with the text, which is based as noted earlier on MAPLE. Also, in most cases the notations used are the same as in the case of MAPLE. If needed, each of these files can be split into separate commands using the "divide cell" option of MATHEMATICA and then evaluated step by step.

M. Symbolic Manipulation Codes – Introduction to MAPLE/MATHEMATICA

Symbolic manipulation codes work with symbols, not only with numbers. As a result, they are capable of performing mathematical operations in an analytical way with great speed and reliability.

Since the early fifties, when the first symbolic codes became available, their development evolved into a scientific discipline essentially based on computer science. Though many such codes are now available, MAPLE, MASCYMA and MATHEMATICA are the most powerful and best-tested among them. A very recent code is AXIOM. Below we present some of the relevant terminology.

Procedure: defines how something is to be performed, the things to do
Arguments: are the things the procedure works with
Primitive: is a built-in procedure
Integers: are whole numbers written without a decimal point. If the point is included, there must be no digits after it.
Floating-point numbers: are those with digits after the decimal point.
Array: is an *n*-dimensional dataset accessible by designating the indices of required items
Assignment: prescribing a value for a variable or changing its value.

M1. MAPLE – first sessions

(MAPLE file vfem1)

A symbolic code used in this book is MAPLE V under the WINDOWS 3.1 operational environment. Most of the files may be processed by older versions of MAPLE too. MAPLE seems to combine the analytical, numerical and graphical capabilities in an optimal proportion. This, together with its simplicity and speed, makes MAPLE an appropriate tool for creating

2 SYMBOLIC MANIPULATION CODES

engineering books and programs. The text below does not intend to replace the MAPLE manuals, but rather stress its features relevant for the following considerations.

Either a semicolon or colon must end a statement. In the latter case MAPLE does not display the output. Starting the line with a hash symbol (#) prevents any evaluation from taking place, and this may be used to temporarily 'turn off' a command or make a comment.

A file may be saved either in the 'user format' or in the 'internal format'. In the later case the filename must end with the extension '.m'. For example, `save 'filename.m'` saves the entire session and `save var1, var2, 'filename.m'` saves the indicated variables in the internal form. The same commands without the extension .m save the file in the 'user format'. Both the interactive (on-line) or batch modes are possible. In the latter case, a file may conveniently be created with the help of Editor and then run by MAPLE. Note, to observe the input and output lines while running a batch the interface should be properly set, for example, as `interface(echo=3);`

To obtain hardcopy use `writeto(filename1);read(filename);` where filename1 is the printed file and filename is a source file. Restarting MAPLE each time one runs a batch file helps to avoid the overlap of assignments and definitions.

Enclosing an expression in a pair of single quotes prevents its evaluation. But each invocation strips off one pair of the quotes,

```
> x:=2:
> y:=2*'x';
```
$$y := 2x$$
```
> y;
```
$$4$$

For the next experience, expand $(a+b)^2$ and then factorize the result:

```
> d1:=expand((a+b)^2);
```
$$d1 := a^2 + 2ab + b^2$$
```
> factor(");
```
$$(a+b)^2$$

To differentiate $\cos x \sin x$ enter

```
> x:='x':d2:=diff(cos(x)*sin(x),x);
```
$$d2 := -\sin(x)^2 + \cos(x)^2$$

Other examples of differentiation are

```
> d3 := diff((a+b)^2,b,b);
    d3 := 2
> d4 := diff((a+b)^2,b$2);
    d4 := 2
> d5 := diff(d1^2,a$3);
    d5 := 24 a + 24 b
```

The code can make substitutions. Say, we need substitute $(a+b)^2$ for z in e^{iz}. Enter

```
> d6:=exp(I*z);d7:=subs(z=(a+b)^2,d6);
    d6 := exp(I z)
    d7 := exp(I(a + b)^2)
```

To assign d7 to, say, g, type

```
> g:=d7;
    g := exp(I(a + b)^2)
```

As noted earlier, the assignment may be removed by a pair of quotes

```
> g:='g';
    g := g
> g;
    g
```

Suppose we need to compute a function $M(L, X) = LX \sin X$ for various L and X. This is conveniently done by a procedure:

```
> M:= proc(L,X)
> L*X*sin(X);
> end;
    M := proc(L,X) L*X*sin(X) end
> M(T,k+F);
    T (k + F) sin(k + F)
```

The command `evalf` (or `evalhf`) may provide a numerical value:

```
> d20:=evalf(M(1,2));
    d20 := 1.818594854
```

4 SYMBOLIC MANIPULATION CODES

```
> Digits:=3;
```
$$\text{Digits} := 3$$
```
> d21:=evalf(M(1,2));
```
$$d21 := 1.82$$
```
> Digits:=6;
```
$$\text{Digits} := 6$$
```
> evalf(M(1,2));
```
$$1.81859$$

where the effect of Digits has also been displayed.
Below we illustrate summation:

```
> x:='x':d22:=sum(a[i]*x^i,i=1..4);
```
$$d22 := a[1]\,x + a[2]\,x^2 + a[3]\,x^3 + a[4]\,x^4$$
```
> y:='y':d23:=sum(sum(b[i,j]*x^i*y^j,i=0..2),j=0..2);
```
$$d23 := b[0,0] + b[1,0]\,x + b[2,0]\,x^2 + b[0,1]\,y + b[1,1]\,x\,y$$
$$+ b[2,1]\,x^2\,y + b[0,2]\,y^2 + b[1,2]\,x\,y^2 + b[2,2]\,x^2\,y^2$$

A pair of quotes enforces the inexplicit form:

```
> d24:='sum(x^i,i=0..3)';
```
$$d24 := \sum_{i=0}^{3} x^i$$
```
> d24;
```
$$1 + x + x^2 + x^3$$

Here is another example of the implicit form:

```
> d25 := 'diff(x^3,x)';
```
$$d25 := \frac{\partial}{\partial x}\,x^3$$
```
> d25;
```
$$3\,x^2$$

This example illustrates a possible evaluation of $\sin 2z$ for $z = \pi/3$:

```
> d27:=sin(2*z);
```
$$d27 := \sin(2\,z)$$

```
> z:=Pi/3;
```
$$z := 1/3\, Pi$$
```
> d27;
```
$$1/2\, 3^{1/2}$$
```
> evalf(");
```
$$.866025$$

and this one illustrates a term selection with the help of `op` and `coeff`:
```
> d28:=a*b*c-v*n+n^2;
```
$$d28 := a\,b\,c - v\,n + n^2$$
```
> op(1,d28);
```
$$a\,b\,c$$
```
> op(2,d28);
```
$$-v\,n$$
```
> coeff(d28,c,1);
```
$$a\,b$$
```
> coeff(d28,n,2);
```
$$1$$

Use `evalc` for operations with complex quantities, such as separation of real and imaginary parts or formulation of conjugate quantity:
```
> d30:=E^(I*f);
```
$$d30 := E^{(I\,f)}$$
```
> evalc(Re(d30));
```
$$\cos(f)$$
```
> evalc(Im(d30));
```
$$\sin(f)$$
```
> evalc(conjugate(d30));
```
$$\cos(f) - \sin(f)\,I$$

The absolute value and polar angle may follow from the `convert` command:

6 SYMBOLIC MANIPULATION CODES

```
> convert(a+b*I,polar);op(1,");
```
$$\text{polar}((a^2 + b^2)^{1/2}, \arctan(b, a))$$
$$(a^2 + b^2)^{1/2}$$

```
> op(1,convert(d30,polar));
```
$$(\cos(f)^2 + \sin(f)^2)^{1/2}$$

A more direct way is to invoke abs:

```
> evalc(abs(d30));
```
$$(\cos(f)^2 + \sin(f)^2)^{1/2}$$

M2. Repetitions, conditionals, expansions and integrations

(MAPLE file vfem1)

A simple example of repetitions is computing $a_i = x^i$ for $i = 2, 4, 6, 8, 10$:

```
> for i from 2 by 2 to 10 do a[i]:=x^i od;
```
$$a[2] := x^2$$
$$a[4] := x^4$$
$$a[6] := x^6$$
$$a[8] := x^8$$
$$a[10] := x^{10}$$

Below we illustrate the use of conditionals by formulating the Heaviside step function:

```
> H:=proc(x)
> if x<0 then 0 else 1 fi;
> end;
```
$$H := \text{proc}(x) \text{ if } x < 0 \text{ then } 0 \text{ else } 1 \text{ fi end}$$

```
> H(-1);
```
$$0$$

```
> H(-0.001);
```
$$0$$

```
> H(0.001);
```
$$1$$
```
> H(1);
```
$$1$$

Here are examples of limits and expansions, which speak for themselves:
```
> limit(sin(x)/x,x=0);
```
$$1$$
```
> limit(x*log(x),x=0,right);
```
$$0$$
```
> limit(x*log(x),x=0,left);
```
$$0$$
```
> taylor(sin(x),x=0,4);
```
$$x - 1/6\,x^3 + O(x^4)$$
```
> asympt(cosh(x),x,2);
```
$$1/2\,\exp(x) + O(\frac{1}{\exp(x)})$$

As to integration, we want to integrate, say, $z/(z^2 + 10)$. Since the previous computations could specify z as a value, first 'unassign' z and then integrate:
```
> z:='z';
```
$$z := z$$
```
> z/(z^2+1);
```
$$\frac{z}{z^2+1}$$
```
> int(',z);
```
$$1/2\,\ln(z^2 + 1)$$

The next example deals with integration of $(\sin x)^2(\cos x)^3$:
```
> x:='x';
```
$$x := x$$
```
> int(sin(x)^2*cos(x)^3,x);
```
$$-1/5\,\sin(x)\cos(x)^4 + 1/15\,\cos(x)\sin(x)^2 + 2/15\,\sin(x)$$

8   SYMBOLIC MANIPULATION CODES

The implicit form of the integral of $x/(b-x)$ and its explicit form are given by

>'int(x/(b-x),x)';

$$\int \frac{x}{b-x}\,dx$$

> ";

$$-x - b\ln(b-x)$$

We have defined earlier the step function of Heaviside, $H(x)$. Try to integrate this from -1 to 1:

> int(H,x=-1..1);

$$\int_{-1}^{1} H\,dx$$

> evalf(");

Error, (in evalf/int) singularity in or near interval of integration

but MAPLE does integrate from 0 to 1:

> int(H,x=0..1);

$$\int_{0}^{1} H\,dx$$

> evalf(");

1.00000

Here are other examples of integration:

> evalf(int(x^2,x=0..1));

.333333

> x:='x':y:='y':z:='z':
> int(int(int(x^3*y*z,x=0..1),y=-1..1),z=0..1);

0

>'int(int(x^2*y,x=0..1),y=0..1)';

$$\int_{0}^{1}\int_{0}^{1} x^2 y\,dx\,dy$$

> evalf(");

.166667

M3. Arrays and matrices

(MAPLE file vfem2)

Arrays enable one to formulate vectors and matrices. A vector is a one-dimensional array, for example:

> u:=array([2,a,s]);

$$u := [2, a, s]$$

> v:=array([[q],[e],[r]]);

$$v := \begin{bmatrix} q \\ e \\ r \end{bmatrix}$$

A matrix is a two-dimensional array:

> b:=array([[2,d,4],[w,t,6],[p,o,y]]);

$$b := \begin{bmatrix} 2 & d & 4 \\ w & t & 6 \\ p & o & y \end{bmatrix}$$

> h:=array([[5,7,10],[x,z,y],[k,L,j]]);

$$h := \begin{bmatrix} 5 & 7 & 10 \\ x & z & y \\ k & L & j \end{bmatrix}$$

Though some of the operations with a matrix may be carried out with the help of evalm, in general one has to invoke the linalg package:

> with(linalg):

   Warning: new definition for norm
   Warning: new definition for trace

> submatrix(h,[1,2],[2,3]);

$$\begin{bmatrix} 7 & 10 \\ z & y \end{bmatrix}$$

> add(b,h);

$$\begin{bmatrix} 7 & d+7 & 14 \\ w+x & t+z & 6+y \\ p+k & o+L & y+j \end{bmatrix}$$

> multiply(b,v);

$$\begin{bmatrix} 2q+de+4r \\ wq+te+6r \\ pq+oe+yr \end{bmatrix}$$

> det(h);

$$5zj - 5yL - 7xj + 10xL + 7ky - 10kz$$

> Q:=inverse(h);

$$Q := \begin{bmatrix} -\dfrac{zj-yL}{\%1} & \dfrac{7j-10L}{\%1} & \dfrac{-7y+10z}{\%1} \\ \dfrac{xj-ky}{\%1} & -5\dfrac{j-2k}{\%1} & -5\dfrac{-y+2x}{\%1} \\ -\dfrac{xL-kz}{\%1} & \dfrac{5L-7k}{\%1} & \dfrac{-5z+7}{\%1} \end{bmatrix}$$

$$\%1 := -5zj + 5yL + 7xj - 10xL - 7ky + 10kz$$

> transpose(v);

$$[q\ e\ r]$$

The operations `mulcol` and `mulrow` enable one to multiply a column or a row by a scalar. The `add` command may also account for multiplication by a scalar, for example:

> mulcol(b,1,Q);

$$\begin{bmatrix} 2Q & d & 4 \\ Qw & t & 6 \\ Qp & o & y \end{bmatrix}$$

> add(b,h,Q,0);

$$\begin{bmatrix} 2Q & Qd & 4Q \\ Qw & Qt & 6Q \\ Qp & Qo & Qy \end{bmatrix}$$

A particularly useful command is `genmatrix`, which states the matrix of coefficients associated with a system of equations:

> eqs:=[a*x + 4*y=9, f*x - e*y=0];

$$eqs := [ax + 4y = 9, fx - ey = 0]$$

> genmatrix(eqs,[x,y]);

$$\begin{bmatrix} a & 4 \\ f & -e \end{bmatrix}$$

The identity array formulates the unit matrix of appropriate size:

> Id:=array(1..2,1..2,identity);

   Id := array(identity, 1..2, 1..2, [])

The need to redefine a matrix entry may frequently arise. Here is an example of redefining b_{23} of the above b matrix as 0 instead of 6:

> b[2,3]:=0;

   b[2, 3] := 0

> print(b);

$$\begin{bmatrix} 2 & d & 4 \\ w & t & 0 \\ p & o & y \end{bmatrix}$$

To formulate a zero matrix one may declare the array as sparse:

> A:=array(sparse,1..2,1..2):print(A);

$$\begin{bmatrix} 0 & 0 \\ 0 & 0 \end{bmatrix}$$

M4. Solving equations

(MAPLE file vfem3)

Solutions to equations may be obtained with the help of solve. Suppose we look for the roots of the equation

$$5q^2 - 7q + 10 = 0$$

Enter

> s:=solve(5*q^2 - 7*q + 10=0,q);

   s := 7/10 + 1/10 I 151$^{1/2}$, 7/10 − 1/10 I 151$^{1/2}$

Each of the roots may be selected by

> s[1];

   7/10 + 1/10 I 151$^{1/2}$

> s[2];

   7/10 − 1/10 I 151$^{1/2}$

Next, solve a system of equations

```
> eq:={a*x[1]+b*x[2]=c,d*x[1]+c*x[2]=f};
    eq := {a x[1] + b x[2] = c, d x[1] + c x[2] = f}
> s1:=solve(eq,{x[1],x[2]});
```

$$s1 := \{x[2] = \frac{af - dc}{ac - db}, \; x[1] = -\frac{bf - c^2}{ac - db}\}$$

```
> s1[1];
```

$$x[2] = \frac{af - dc}{ac - db}$$

```
> s1[2];
```

$$x[1] = -\frac{bf - c^2}{ac - db}$$

If we need the values of the roots only, we may use `assign`:

```
> assign(s1);
```

Then to invoke, say, the above x_2 value, enter

```
> x[2];
```

$$\frac{af - dc}{ac - db}$$

Alternatively, a linear system of equations may be solved with the help of the `linalg` package and resorting to `linsolve`. To solve the system

$$ax_1 + bx_2 = c$$
$$dx_1 + ex_2 = f$$

type in

```
> A:=array([[a,b],[d,e]]);
```

$$A := \begin{bmatrix} a & b \\ d & e \end{bmatrix}$$

```
> B:=array([c,f]);
    B := [c, f]
> with(linalg,linsolve):
> linsolve(A,B);
```

$$[-\frac{bf - ce}{ae - db}, \frac{af - dc}{ae - db}]$$

The dsolve command may yield solutions to differential equations. Here is an example of solving the classical oscillator equation:

$$mx_{,tt} + kx = 0$$

```
> x:='x';
```

$$x := x$$

```
> eq:=m*diff(x(t),t,t) + k*x(t);
```

$$eq := m\left(\frac{d^2}{dt^2}x(t)\right) + k\,x(t)$$

```
> dsolve(eq,x(t));
```

$$x(t) = \_C1\,\exp\left(\frac{Ik^{1/2}t}{m^{1/2}}\right) + \_C2\,\exp\left(\frac{Ik^{1/2}t}{m^{1/2}}\right)$$

where _C1 and _C2 are free constants to be specified from initial (boundary) conditions. Proper commands are also available in the MAPLE library.

M5. Hints

(MAPLE file vfem4)

The op command helps to extract operands of an expression. Section M1 contains some examples. This command is particularly useful for matrix operations. Consider three matrices n, f and g, where n is a product of f and g, that is, $n = f.g$:

```
> n:=array([[a],[b]]);
```

$$n := \begin{bmatrix} a \\ b \end{bmatrix}$$

```
> f:=array([[1,3],[2,4]]);
```

$$f := \begin{bmatrix} 1 & 3 \\ 2 & 4 \end{bmatrix}$$

```
> g:=array([[x],[y]]);
```

$$g := \begin{bmatrix} x \\ y \end{bmatrix}$$

```
> with(linalg,multiply):
```

14 SYMBOLIC MANIPULATION CODES

```
> n=multiply(f,g);
```

$$n = \begin{bmatrix} x+3y \\ 2x+4y \end{bmatrix}$$

A visually more informative form follows from

```
> op(n)=multiply(f,g);
```

$$\begin{bmatrix} a \\ b \end{bmatrix} = \begin{bmatrix} x+3y \\ 2x+4y \end{bmatrix}$$

Another example involves substitutions. The subs command alone may not work in case of a matrix and the invocation of op is needed:

```
> subs({a=1,b=8},n);

    n

> subs({a=1,b=8},op(n));
```

$$\begin{bmatrix} 1 \\ 8 \end{bmatrix}$$

Note, the original form of *n* has not changed:

```
> print(n);
```

$$\begin{bmatrix} a \\ b \end{bmatrix}$$

Another way to perform substitutions is to use the do command

```
> for i to 2 do m[i,1]:=subs({a=1,b=8},n[i,1]) od;

    m[1,1] := 1
    m[2,1] := 8
```

which yields a new *m* array from the *n* array.

The map function provides a convenient way of differentiating or integrating matrices:

```
> map(diff,op(n),a);
```

$$\begin{bmatrix} 1 \\ 0 \end{bmatrix}$$

```
> map(int,op(n),a=1..5);
```

$$\begin{bmatrix} 12 \\ 4b \end{bmatrix}$$

Alternatively, the same operations may be performed with the help of do.

Further, a need to create a sequence of expressions often arises in computation. For example, this could be a sequence of unknown quantities, for which a system of equations must be resolved. Here are two convenient ways of creating a sequence:

```
> A[j]$ j=3..8;
```
$\qquad$ A[3], A[4], A[5], A[6], A[7], A[8]
```
> A.(3..8);
```
$\qquad$ A3, A4, A5, A6, A7, A8

Procedures provide a convenient means for automated computations. Suppose we need derivatives of various orders of various functions with respect to t. Define P as follows:

```
> P:=proc(T,n)
> diff(T,t$n):
> end:
```

Then, if we need to differentiate the expression $t^2 \sin t$ twice, enter

```
> P(sin(t)*t^2,2);
```
$\qquad - \sin(t) t^2 + 4 \cos(t) t + 2 \sin(t)$

M6. MATHEMATICA – first sessions

(MATHEMATICA file vfem1)

MATHEMATICA labels the input and output lines unlike MAPLE. This is a feature it shares with MACSYMA. Another major difference is the use of brackets: ( ) for grouping terms, [ ] for argument of functions, { } for lists, [[ ]] for indexing. Lists are enclosed in curly braces { } unlike the square brackets [ ] of MAPLE. Note that all MATHEMATICA built-in names begin with a capital letter. In particular, N is used for a numerical evaluation of the arguments. It is therefore a good habit for users to start their own notation with a lower-case letter.

Here are the elementary notations of MATHEMATICA:

$x =$ assigning a value to x,
$x =.$ removing the value,
; supressing the display of the output,
% referring to the immediate result above.

Below are computations with MATHEMATICA similar to those of Section M1. The first ident indicates the input and the second ident the output. The latter also has a different font.

16 SYMBOLIC MANIPULATION CODES

```
x=10
```
  10
```
x
```
  10
```
x=.
x
```
  x

Multiplication does not necessarily require the *-sign

```
y=6
```
  6
```
2*y
```
  12
```
2y
2 y
```
  12
  12

Here are operations with lists:

```
a={19,3,b}
a
```
  {19, 3, b}
  {19, 3, b}
```
g={1,4,3}
a+g
```
  {1, 4, 3}
  {20, 7, 3 + b}

Picking up a term of g:

```
Part[g,2]
g[[2]]
```
  4
  4

Changing the first term in list

```
a[[1]]=B
```
  B

a

{B, 3, b}

Here are examples of differentiation:

D[Cos[x]Sin[x]]

$\cos^2[x] - \sin^2[x]$

D[Cos[x]Sin[x],{x,2}]

$-4 \cos[x] \sin[x]$

and substitution $z = (a+b)^2$ in exp(iz), where the value of a has been removed by a=.

d6=E^(I z)/.z->(a+b)^2

$E^{I(b+a)^2}$

For automated computations one may use a function, which is similar to the procedure of **MAPLE**. For example, the function below computes $LX \sin X$ for various L and X:

f[L_, X_]:=L X Sin[X]
f[T,k+F]

$(F+k) T \sin[F+k]$

which is now evaluated numerically for $T = 2$, $k = 3$ and $F = 4$:

N[%]/.{T->2,k->3,F->4}

14 Sin[7]

This function can be removed by

Clear[f];

Here are examples of summation:

Sum[a[i] x^i,{i,1,4}]

$x \, a[1] + x^2 a[2] + x^3 a[3] + x^4 a[4]$

and complex numbers

d30=Exp[5 I]
Re[d30]
Im[d30]

E^{5I}

Cos[5]

Sin[5]

18 SYMBOLIC MANIPULATION CODES

In order to work with complex-valued symbols one must load the package Algebra 'ReIm' as follows:

```
<<Algebra'ReIm'
```

Consider the example,

```
d30=E^(I*f)
```

Declare first that the above *f* is real and find the real and imaginary parts:

```
Im[f]^=0
Re[d30]
Im[d30]
```

$$E^{I\,f}$$
$$0$$
Cos[f]
Sin[f]

The repetitions follow from the do command:

```
a[1]=2;a[2]=1;
Do[Print[a[i]],{i,1,2}]
```

2
1

and conditionals make use of If

```
H[x_]:=If[x<0,0,1]
H[-1]
H[-0.01]
H[0.01]
```

0
0
1

Now we illustrate limits and expansions:

```
Limit[Sin[x]/x,x->0]
Limit[x Log[x],x->0]
Series[Sin[x],{x,0,4}]
```

1
0
$$x - \frac{x^3}{6} + 0[x]^5$$

Finally, here are examples of integrations:

Integrate[x/(x^2+1),x]

$$\frac{Log[1+x^2]}{2}$$

NIntegrate[H[x],{x,-1,1}]

1.

The associated MATHEMATICA file contains other examples too.

M7. Tables and matrices

(MATHEMATICA file vfem2)

Here are examples of tables and matrices in MATHEMATICA:

u=Table[a[i],{i,2,10,2}]

{a[2], a[4], a[6], a[8], a[10]}

TableForm[%]

a[2]
a[4]
a[6]
a[8]
a[10]

Note the difference in presentation of the above two versions of u. More examples:

v={q,e,r}
b={{2,d,4},{w,t,6},{p,o,y}}
b[[2,3]]=9
MatrixForm[b]

{q, e, r}
{{2, d, 4}, {w, t, 6}, {p, o, y}}
9
2 d 4
w t 9
p o y

Note that MATHEMATICA does not provide brackets for a two-dimensional presentation of a matrix and that b_{23} in matrix b received a new value.

20 SYMBOLIC MANIPULATION CODES

Here are the two versions of the dot product:

```
s=b.v
Dot[b,v]
```

$\{de + 2q + 4r, 9r + et + qw, eo + pq + ry\}$
$\{de + 2q + 4r, 9r + et + qw, eo + pq + ry\}$

MATHEMATICA does not differ between 'rows' and 'columns':

```
v.b
```

$\{2q + pr + ew, dq + or + et, 9e + 4q + ry\}$

```
b.b
```

$\{\{4 + 4p + dw, 2d + 4o + dt, 8 + 9d + 4y\},$
$\{9p + 2w + tw, 9o + t^2 + dw, 9t + 4w + 9y\},$
$\{2p + ow + py, dp + ot + oy, 9o + 4p + y^2\}\}$

```
b+b
```

$\{\{4, 2d, 8\}, \{2w, 2t, 18\}, \{2p, 2o, 2y\}\}$

Below s is a submatrix of h:

```
h={{5, 7, 10},{x, z, y},{k, L, j}}
s=h[[{1,2},{2,3}]]
```

$\{\{5, 7, 10\}, \{x, z, y\}, \{k, L, j\}\}$
$\{\{7, 10\}, \{z, y\}\}$

The basic code of MATHEMATICA does not seem to have a command similar to genmatrix of MAPLE. Assume that a system of equations is given by the product of matrices n1 and n2 spesified below. Then the program below extracts matrix n1 from the dot product of n1 and n2:

```
n1={{a1,a2},{a3,a4}};n2={x[1],x[2]};n=Dot[n1,n2]
K=Array[m,{2,2}]
Do[m[i,j]=Coefficient[n[[i]],x[j]],{i,1,2},{j,1,2}]
MatrixForm[K]
```

$\{a1\,x[1] + a2\,x[2], a3\,x[1] + a4\,x[2]\}$
$\{\{m[1, 1], m[1, 2]\}, \{m[2, 1], m[2, 2]\}\}$
a1 a2
a3 a4

M8. Solving equations

(MATHEMATICA file vfem3)

Consider a solution to an algebraic equation:

```
s=Solve[5*q^2-7*q+10==0,q]
```

$$\{\{q \to \frac{7 - I\,\text{Sqrt}[151]}{10}\}, \{q \to \frac{7 + I\,\text{Sqrt}[151]}{10}\}\}$$

and to a system of equations

```
eq={a*x[1]+b*x[2]==c,d*x[1]+c*x[2]==f}
s1=Solve[eq,{x[1],x[2]}]
```

$$\{a\,x[1] + b\,x[2] == c, d\,x[1] + c\,x[2] == f\}$$

$$\{\{x[1] \to \frac{c}{a} + \frac{b\,(c\,d - a\,f)}{a\,(a\,c - b\,d)}, x[2] \to -(\frac{c\,d - a\,f}{a\,c - b\,d})\}\}$$

The roots therefore take the values

```
x[1]/.%
x[2]/.%%
```

$$\{\frac{c}{a} + \frac{b\,(c\,d - a\,f)}{a\,(a\,c - b\,d)}\}$$

$$\{-(\frac{c\,d - a\,f}{a\,c - b\,d})\}$$

Here is a solution in a matrix form:

```
A={{a,b},{d,e}}
B={c,f}
LinearSolve[A,B]
```

$$\{\{a, b\}, \{d, e\}\}$$
$$\{c, f\}$$

$$\{\frac{c}{a} - \frac{b\,(-(c\,d) + a\,f)}{a\,(-(b\,d) + a\,e)}, \frac{-(c\,d) + a\,f}{-(b\,d) + a\,e}\}$$

$$\{\frac{-(c\,e) + (b\,f)}{b\,d - a\,e}, \frac{-(c\,d) + a\,f}{-(b\,d) + a\,e}\}$$

Consider a system of differential equations:

```
eq1=z'[t]-p[t]==0
eq2=m*p'[t]+k*z[t]==0
```

$$-p[t] + z'[t] == 0$$
$$k\,z[t] + m\,p'[t] == 0$$

The solution is given by

```
DSolve[{eq1,eq2},{z[t],p[t]},t];
```

where the display of lengthy expressions has been prevented. The function $z(t)$, for example, is given by

```
z[t]/.%;
Simplify[%]
```

$$\{(E^{(-I\,Sqrt[k]\,t)/Sqrt[m]}\,(Sqrt[k]\,C[1] + E^{(2I\,Sqrt[k]\,t)/Sqrt[m]}\,Sqrt[k]\,C[1] + I\,Sqrt[m]\,C[2] - I\,E^{(2I\,Sqrt[k]\,t)/Sqrt[m]}\,Sqrt[m]\,C[2]))/(2\,Sqrt[k])\}$$

M9. Hints

(MATHEMATICA file vfem4)

MATHEMATICA has a covenient means for local substitution, particularly, in matrices. Say $a = 2$ and $d = \sin x$ and this must be substituted in matrix n,

```
n=MatrixForm[{{a,b},{c,d}}]
n/.{a → 2, d → Sin[x]}
```

$$\begin{matrix} a & b \\ c & d \\ 2 & b \\ c & Sin[x] \end{matrix}$$

Here are examples of differentiation and integration of a matrix:

```
p=%[[]]
D[p,x]
Integrate[p,x]
```

$$\{\{2, b\}, \{c, Sin[x]\}\}$$
$$\{\{0, 0\}, \{0, Cos[x]\}\}$$
$$\{\{2x, bx\}, \{cx, -Cos[x]\}\}$$

The following routine may be used for constructing a sequence

```
Table[a[i],{i,0,10,2}]
```

$$\{a[0], a[2], a[4], a[6], a[8], a[10]\}$$

or a more complex version convenient for assigning values to a sequence of variables,

```
For[i=1,i<=2,i++,{a[i]=2*i,b[i]=a[i]+1}];
a[1]
b[1]
```

    2
    3

```
a[2]
b[2]
```

    4
    5

```
a[3]
b[3]
```

    a[3]
    b[3]

For automated computations one may use the function procedure. Say, to compute the nth derivative of $T(t)$, set

```
f[T_,n_]:=D[T,{tt,n}];
f[Sin[tt]*tt^2,2]
```

    $4\,tt\,\cos[tt] + 2\,\sin[tt] - tt^2\,\sin[tt]$

D. Direct Methods

The exact definition of direct methods of engineering analysis is difficult to formulate. These are mainly approximate methods, which reduce the mathematical complexity by an appeal to additional information available on the nature of the solution. They thus break with a predominantly Western tradition, which seems to give priority to the exact solution, regardless of the circumstances. The difference between the purely mathematical and engineering approaches manifests itself in direct methods. The latter take full advantage of the fact that the need for an exact solution is a rarity in engineering. From this viewpoint, a sufficiently accurate approximate solution may be as satisfactory as any other. The finite element technique is an outgrowth of these methods. The use of symbolic computation allows for a relatively easy evaluation of higher approximations, which has traditionally been a difficulty under the conventional performance.

As noted earlier, in engineering we look for a sufficiently accurate solution rather than for the exact one. Since the very formulation of an engineering problem contains a degree of uncertainty, the pursuit of the exact solution may not be justified, particularly if it is a time-consuming and complicated one.

The local behaviour of the function of interest may be of secondary relevance from a practical point of view, which may stress the various relevant functionals instead. Consequently, an approximate solution, which differs locally from the exact one but provides a satisfactory prediction for the functionals, may be acceptable.

As far as the closeness between two functions u_1 and u_2 is concerned, introduce the quantity δ by

$$\delta^2 = \int_V (u_1 - u_2)^2 \, dv \tag{D.1}$$

The smaller δ, the closer u_1 and u_2 are in the *mean*. Note, this type of closeness says nothing about the local proximity of the functions. In looking for approximate solutions, we think, first of all, of their closeness to the exact one in the mean, which usually constitutes a much simpler problem than a search for local approximations. This may be sufficient for engineering purposes.

The idea of direct methods was originated by Rayleigh and then extended by Ritz. Since then various techniques have been devised, all of which may be thought of as modifications of that idea.

This chapter begins with a consideration of the basic concepts of variational calculus. Among the variety of direct methods, we treat the Bubnov–Galerkin and Rayleigh–Ritz techniques, which are the most basic ones. We prefer to introduce these in an intuitive way suitable for engineers and only then consider relevant mathematical aspects.

An overview of the weighted residual approach is also given, as well as the iterative Kantorovich method. The latter may require involved computations, but is highly accurate. It is particularly suitable for realization via a symbolic manipulation code.

In general, the presentation refers to results obtained via *vfem* files with the extension '.ms' (MAPLE). To derive the relevant results with the help of MATHEMATICA the user must run the *vfem* files with the extension '.ma'.

D1. The functional and its stationarity

The specification of unknown functions or parameters, which is the major objective of engineering analysis, may be achieved, among others, by investigating the extrema of certain quantities referred to as functionals. This has proved particularly useful for the solution of practically oriented problems.

Consider a function $y(x)$ defined for the closed interval $[x_1, x_2]$ and satisfying the boundary conditions

$$y(x_1) = y_1, \qquad y(x_2) = y_2 \tag{D1.1}$$

and introduce the integral of interest

$$J(y) = \int_{x_1}^{x_2} f(x, y, y_{,x}) \, dx \tag{D1.2}$$

In (D1.2) $f(x, y, y_{,x})$ is thought of as a given function of the indicated arguments, while the function $y(x)$ may be chosen from a set of *admissible* functions $y_i(x)$, $i = 1, 2, \ldots$. On substitution of $y_i(x)$ in (D1.2), one arrives at a number, $J(y_i)$,

$$J(y_i) = \int_{x_1}^{x_2} f(x, y_i, y_{i,x}) \, dx \tag{D1.3}$$

THE FUNCTIONAL AND ITS STATIONARITY 27

In this sense (D1.3) represents a scalar-valued function of $y(x)$ and the expression (D1.2) is said to be a *functional*. Generally speaking, a correspondence between a function and a scalar may be viewed as a functional.

A question arises about that particular function $y_0(x)$ among the above set of admissible functions, which would provide an extremum to $J(y)$. To this end, suppose that each $y_i(x)$ admits a representation

$$y_i(x) = y_0(x) + \epsilon q_i(x) \tag{D1.4}$$

where $q(x)$ is a continuous function and ϵ is a small parameter (Fig. D1).

To meet the boundary conditions (D1.1) we must set

$$q(x_1) = 0, \quad q(x_2) = 0 \tag{D1.5}$$

On substitution of (D1.4) in (D1.3), the latter becomes a function of ϵ and for the extremum one may write

$$dJ/d\epsilon = 0 \quad \text{for} \quad \epsilon = 0 \tag{D1.6}$$

The condition $\epsilon = 0$ in (D1.6) stems from (D1.4), which shows that the *extremizer* $y_0(x)$ follows, if ϵ vanishes.

The small parameter ϵ enters the functional J through y and $y_{,x}$. Therefore,

$$dJ/d\epsilon = \int_{x_1}^{x_2} [(\partial f/\partial y)(\partial y/\partial \epsilon) + (\partial f/\partial y_{,x})(\partial y_{,x}/\partial \epsilon)] \, dx$$

$$= \int_{x_1}^{x_2} [(\partial f/\partial y)q + (\partial f/\partial y_{,x})q_{,x}] \, dx = 0 \tag{D1.7}$$

where use has been made of (D1.4). Integrating by parts the second term, we get

$$\int_{x_1}^{x_2} (\partial f/\partial y_{,x})q_{,x} dx = (\partial f/\partial y_{,x})q \Big|_{x_1}^{x_2} - \int_{x_1}^{x_2} [d(\partial f/\partial y_{,x})/dx]q \, dx \tag{D1.8}$$

where the conditions (D1.5), namely, $q(x_1) = 0$, $q(x_2) = 0$, have been incorporated. The basic equation (D1.7) becomes

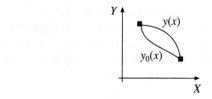

D1 Admissible functions

$$\int_{x_1}^{x_2} [\partial f/\partial y - d(\partial f/\partial y_{,x})/dx] q \, dx = 0 \qquad (D1.9)$$

Since $q(x)$ in (D1.9) is an arbitrary function, this expression suggests that the bracketed term must vanish. Without pursuing a proof, we note that this conclusion indeed holds under certain conditions of regularity imposed on $y(x)$, $q(x)$ and $f(x, y, y_{,x})$. Thus, we get the differential equation which the extremizer $y_0(x)$ must obey:

$$\partial f/\partial y - d(\partial f/\partial y_{,x})/dx = 0 \qquad (D1.10)$$

This is referred to as the Euler–Lagrange equation of the functional (D1.2).

The derivation has been based on the assumption that $y_0(x)$ is an extremizer, and the Euler–Lagrange equation (D1.10) is therefore a necessary condition only. Also, the nature of extremum has remained unspecified. The associated value of the functional, $J(y_0)$, is said to be stationary. The sign of the second derivative $J_{,\epsilon\epsilon}$ may indicate this nature, with $J_{,\epsilon\epsilon} > 0$ indicating a minimum and $J_{,\epsilon\epsilon} < 0$ a maximum.

The derivative of a functional with respect to ϵ evaluted at $\epsilon = 0$ has been the core of the above considerations and is worthy of a special notation. Here it is

$$\frac{dJ}{d\epsilon}(\epsilon = 0)\epsilon = \delta J \qquad (D1.11)$$

which is referred to as the first variation of the functional J. Consequently, the stationarity condition takes the form

$$\delta J(y) = \int_{x_1}^{x_2} \delta f(x, y, y_{,x}) dx = 0 \qquad (D1.12)$$

where neither the independent variable x nor the limits of integration are subjected to variations.

D2. Hamilton's principle

Hamilton's principle provides an excellent illustration of a generic postulate capable of describing a variety of particular cases. Begin with a system of particles and assume that on the observation time interval (t_1, t_2) their positions are given by the generalized coordinates $\eta_j(t), j = 1, 2, \ldots, n$, which have continuous derivatives up to the second order, inclusively, and satisfy the boundary conditions

$$\eta_j(t_1) = \eta_{j1}, \qquad \eta_j(t_2) = \eta_{j2} \qquad (D2.1)$$

A function satisfying the above requirements is an admissible function. Assuming that the potential energy of the system U does exist and is a function of $\eta_j(t)$, and that the kinetic energy T depends on $\eta_j(t)$ and its derivative $\eta_{j,t}(t)$, we state Hamilton's principle as the condition of stationarity of the following functional:

$$I = \int_{t_1}^{t_2} (T - U)\, dt \qquad (D2.2)$$

Namely, the above functional becomes stationary for the actual motion of the system in comparison with neighbouring admissible motions. If this is so, the Euler–Lagrange equations of (D2.2) should provide the equations of motion.

The first variation of (D2.2) is

$$\delta I = 0 \qquad (D2.3)$$

and for the admissible motion $\eta_j(t)$, (D1.4) yields

$$\eta_j(t) = \eta_j^0(t) + \epsilon q_j(t) \qquad (D2.4)$$

Here $\eta_j^0(t)$, $j = 1, 2, \ldots, n$ describe the actual motion (the extremizers, according to Hamilton's principle). Note, $q_j(t)$ must vanish at $t = t_1$ and $t = t_2$ to obey the boundary conditions (D2.1). Equations (D2.3), (D2.4) and (D2.2) provide

$$\delta I = \int_{t_1}^{t_2} \left[(\partial T/\partial \eta_j) q_j + (\partial T/\partial \eta_{j,t}) q_{j,t} - (\partial U/\partial \eta_j) q_j \right] dt \qquad (D2.5)$$

where the repeated index j implies summation from 1 to n. It has been assumed that U depends on the coordinates η_j and T depends on the coordinates η_j and its first derivatives. Integrating, as in Section D1, the second term in (D2.5) by parts, taking into account (D2.1) and introducing the Lagrangian

$$L = T - U \qquad (D2.6)$$

we get from (D2.5)

$$\delta I = \int_{t_1}^{t_2} [\partial L/\partial \eta_j - d(\partial L/\partial \eta_{j,t})/dt] q_j\, dt = 0 \qquad (D2.7)$$

This, in view of the above regularity conditions, yields a result similar to (D1.10):

$$\partial L/\partial \eta_j - d(\partial L/\partial \eta_{j,t})/dt = 0 \qquad (D2.8)$$

Note that (D2.8) holds for any of the functions $\eta_j^0(t)$, $j = 1, 2, \ldots, n$. Therefore this is a system of equations in a number equal to the number of degrees of freedom.

For an example of the harmonic oscillator, let

$$T = mx_{,tt}^2/2, \quad U = kx^2/2, \quad L = mx_{,tt}^2/2 - kx^2/2 \tag{D2.9}$$

with $x(t)$ the coordinate and k the stiffness. Equation (D2.8) yields the well-known result

$$mx_{,tt} + kx = 0 \tag{D2.10}$$

As to continuous systems, one may introduce the Lagrangian density L_0 by

$$L_0 = T_0 - U_0 \tag{D2.11}$$

where T_0 and U_0 are the specific kinetic and potential energies, respectively. These quantities depend on a set of functions η_i, $i = 1, 2, \ldots, k$, and their temporal and spatial derivatives. For example, in the case of elastic systems there are three functions of displacement η_i, $i = 1, 2, 3$. Hamilton's principle postulates

$$\delta I = \int_{t_1}^{t_2} \int_V \delta L_0 \, dv \, dt = 0 \tag{D2.12}$$

where V is the relevant volume and $dv = dx_1 dx_2 dx_3$. This is similar to (D2.3).

The evaluation of (D2.12) resembles that of (D2.3). In particular, the variation from a technical viewpoint may be considered as a differential. Therefore, if the Lagrangian density L_0 depends on η_i, $i = 1, 2, \ldots, k$, and its first derivatives only, this yields for any of η_i, $i = 1, 2, \ldots, k$ (summation with respect to $j = 1, 2, 3$ applies):

$$\delta L_0 = (\partial L_0/\partial \eta)\delta \eta + (\partial L_0/\partial \eta_{,t})\delta \eta_{,t} + (\partial L_0/\partial \eta_{,j})\delta \eta_{,j} \tag{D2.13}$$

Substituting this in (D2.12) and integrating by parts in a way similar to that of Section D1, we obtain

$$\int_{t_1}^{t_2} \int_V [d(\partial L_0/\partial \eta_{,t})/dt + d(\partial L_0/\partial \eta_{,j})/dx_j - \partial L_0/\partial \eta]\delta \eta \, dv \, dt = 0 \tag{D2.14}$$

Therefore, the Euler–Lagrange equation for each of η_i, $i = 1, 2, \ldots, k$, is

$$d(\partial L_0/\partial \eta_{,t})/dt + d(\partial L_0/\partial \eta_{,j})/dx_j - \partial L_0/\partial \eta = 0 \tag{D2.15}$$

Note a remarkable similarity in treating the spatial and temporal derivatives, which (D2.15) shows. This approach generalizes to the case when L_0 depends also on the second or higher spatial derivatives of η_i, $i = 1, 2, \ldots, k$.

D3. Minimum of potential energy

As the above results indicate, governing equations may follow from a single variational principle as a necessary condition of its stationarity. It can be shown that for static conservative systems the condition of stationarity is also the condition for a minimum, provided the equilibrium state is stable. Fig. D2 illustrates qualitatively the difference between a stable and unstable equilibria. Moreover, a type of boundary condition, the so-called natural conditions, may also follow from a variational principle.

As an example, consider an elastic uniform cantilever subjected to a static force P at its end $x = L$. The boundary conditions for the deflection $w^0(x)$ are

$$w^0(x=0) = w^0_{,x}(x=0) = 0 \tag{D3.1}$$

The potential energy is

$$U = EI/2 \int_0^L (w^0_{,xx})^2 dx - Pw^0(x=L) \tag{D3.2}$$

Represent an admissible deflection, as earlier, by

$$w(x) = w^0(x) + \epsilon q(x) \tag{D3.3}$$

and find the associated value of the functional (D3.2)

$$U + \Delta U = EI/2 \int_0^L (w^0_{,xx} + \epsilon q_{,xx})^2 dx - P[w^0(x=L) + \epsilon q(x=L)] \tag{D3.4}$$

This, in view of (D3.2), yields the increment ΔU:

$$\Delta U = EI/2 \int_0^L (2\epsilon w^0_{,xx} q_{,xx} + \epsilon^2 q_{,xx}^2) dx - P\epsilon q(x=L) \tag{D3.5}$$

This expression contains terms of the first and second order of smallness, namely,

$$\Delta U = \delta U + \delta^2 U/2 \tag{D3.6}$$

b) unstable equilibrium

a) stable equilibrium

D2 Illustrating stable and unstable equilibria

where

$$\delta U = \epsilon \left[EI \int_0^L w^0_{,xx} q_{,xx}\, dx - Pq(x=L) \right] \tag{D3.7}$$

$$\delta^2 U = \epsilon^2 EI \int_0^L q_{,xx}^2\, dx \tag{D3.8}$$

Since $\delta^2 U$, the so-called *second* variation, is obviously positive, the vanishing first variation, $\delta U = 0$, provides a local minimum of U.

In order to complete the analysis, employ again integration by parts (twice). Then (D3.7) yields

$$\begin{aligned}\delta U = \epsilon EI \bigg[&\int_0^L w^0_{,xxxx} q\, dx - w^0_{,xxx}(x=L) q(x=L) \\ &+ w^0_{,xxx}(x=0) q(x=0) + w^0_{,xx}(x=L) q_{,x}(x=L) \\ &- w^0_{,xx}(x=0) q_{,x}(x=0) \bigg] - \epsilon Pq(x=L)\end{aligned} \tag{D3.9}$$

Because of the boundary conditions (D3.1), $q(x=0) = q_{,x}(x=0) = 0$ and (D3.9) simplifies to

$$\begin{aligned}\delta U = \epsilon EI &\left[\int_0^L w^0_{,xxxx} q\, dx + w^0_{,xx}(x=L) q_{,x}(x=L) \right] \\ &- \epsilon[Pq(x=L) + EI w^0_{,xxx}(x=L) q(x=L)]\end{aligned} \tag{D3.10}$$

For an arbitrary $q(x)$ (but satisfying the above conditions $q(x=0) = q_{,x}(x=0) = 0$), δU vanishes, if

$$w^0_{,xxxx} = 0 \tag{D3.11}$$

and

$$w^0_{,xx}(x=L) = 0, \qquad P + EI w^0_{,xxx}(x=L) = 0 \tag{D3.12}$$

Equation (D3.11) is the Euler–Lagrange equation of the functional U given by (D3.2). Equations (D3.12) are the moment and shear equations for the free end, respectively. The expressions (D3.12) are referred to as the *natural* boundary conditions, as they follow from the stationarity of the functional. On the other hand, the boundary conditions (D3.1) are the *essential* (imposed) ones.

D4. The Bubnov–Galerkin method

Consider a simple example of the equation governing a beam on an elastic foundation:

$$EIw_{,xxxx} + Kw - q = 0 \tag{D4.1}$$

where w is the deflection, EI the beam stiffness, K the foundation coefficient and q the intensity of distributed load. Let the beam be clamped, which implies

$$w(x=0) = w(x=L) = 0$$
$$w_{,x}(x=0) = w_{,x}(x=L) = 0 \tag{D4.2}$$

with L being the beam length.

Even though this boundary value problem may be solved exactly, we look for an approximate solution. Note that, if this solution satisfies (D4.2), its substitution in (D4.1) yields a nonvanishing residual. Then a minimization of this residual may provide an approximate solution to the problem stated by (D4.1) and (D4.2).

Consider the so-called *trial* functions

$$a(L-x)^2 x^2, \quad b(L-x)^2 x^3, \ldots \tag{D4.3}$$

with a and b being unspecified coefficients. These satisfy (D4.2). Therefore w may be expanded as

$$w(x) = (L-x)^2(ax^2 + bx^3 + \cdots) \tag{D4.4}$$

and substituted in (D4.1) to yield a residual $f(a,b,q,x)$. Note: (i) the coefficients a and b are at our disposal to specify them so as to minimize the 'error' in satisfying (D4.1), (ii) in the expansion (D4.4) we preserve as low powers of x as the boundary conditions (D4.2) allow for.

This minimization may be performed in a variety of ways, and one of them consists of resorting to Hamilton's principle. Recall the basic result of Section D2, dealing with Hamilton's principle, namely (D2.14)

$$\int_{t_1}^{t_2}\int_V [\mathrm{d}(\partial L_0/\partial \eta_{,t})/\mathrm{d}t + \mathrm{d}(\partial L_0/\partial \eta_{,j})/\mathrm{d}x_j - \partial L_0/\partial \eta]\delta\eta \, \mathrm{d}v \, \mathrm{d}t = 0 \tag{D4.5}$$

This equation states that if η is the *exact* solution, then the above integral must vanish for an *arbitrary* $\delta\eta$. Instead, looking for an *approximate* solution, one may require this integral to vanish for a specifically defined $\delta\eta$, for example, if $\delta\eta$ is a variation of the trial functions.

Considering η as the deflection w and the bracketed term in (D4.5) as the left-hand side of (D4.1), we get from (D4.5)

$$\int_0^L f(a,b,q,x)\delta w \, \mathrm{d}x = 0 \tag{D4.6}$$

where the limits of integration have been properly adjusted and $f(a,b,q,x)$ is the residual. It follows from (D4.4) that

$$\delta w(x) = (L-x)^2(\delta a x^2 + \delta b x^3) \tag{D4.7}$$

and (D4.6) provides

$$\int_0^L f(a,b,q,x)(L-x)^2(x^2\delta a + x^3\delta b)\,dx = 0 \tag{D4.8}$$

This yields two algebraic equations for the two unknown coefficients a and b;

$$\int_0^L f(a,b,q,x)(L-x)^2 x^2\,dx = 0 \tag{D4.9}$$

and

$$\int_0^L f(a,b,q,x)(L-x)^2 x^3\,dx = 0 \tag{D4.10}$$

Resolving (D4.9) and (D4.10) for a and b and substituting these in (D4.4) one gets an approximation of the *second* order (two free coefficients) to the deflection w. On the other hand, setting from the beginning $b=0$ in (D4.4) would provide a single equation for the single unknown coefficient a, which is the first approximation. Similarly, continuing (D4.4) so as to include higher terms, one could derive a more accurate approximation. With a new coefficient, one would always get a new algebraic equation, preserving thereby the necessary correspondence.

The choice of trial functions, such as (D4.3), has a crucial influence on the accuracy. These functions, first of all, must meet boundary conditions. Besides power series, a trigonometric series may be employed too. In doing so, the lowest possible frequencies (powers) must be preserved. Recalling the fact that a sufficiently regular function admits expansion in a Taylor or Fourier series, one may better appreciate these comments.

In the above we thought of direct methods in terms of a subsequent approximation of the unknown function with the help of a series expansion. Alternatively, these methods allow for an approximate representation of a continuous system, such as (D4.1), by a discrete one, such as (D4.4). Indeed, the representation (D4.4) is one of a system of two degrees of freedom (two unspecified coefficients). The search for a higher approximation now appears as a better modelling of a continuous system by a discrete one.

Equations (D4.9) and (D4.10) represent a particular case of the so-called *weighted residual* approach, as they consist of a residual multiplied by a weight. This approach gives rise to a variety of methods similar to the present one, but based on other weight functions (see section D14).

D5. Beam on elastic foundation

(MAPLE file vfem18)

Turning to computations, formulate the equation (D4.1) and then, looking for the first approximation, a trial function $A(x - L)^2 x^2$, as given by (D4.3),

```
> d1 := diff(EI*w(x),x$4)+K*w(x) - q:
> d2 := A*(L-x)^2*x^2:
```

Substituting the trial function in the equation, arrive at the residual f, then multiply this by the trial function and integrate from 0 to L, according to (D4.9):

```
> f := subs(w(x)=d2,d1):"*d2/A:int(",x=0..L):
```

It remains to find A from this equation and substitute the result in d2

```
> d4 := solve(",A):simplify(d4);
```

$$21 \frac{q}{KL^4 + 504 EI} \qquad (D5.1)$$

```
> deflection := subs(A=",d2);
```

$$\text{deflection} := 21 \frac{q(L-x)^2 x^2}{KL^4 + 504 EI}$$

An alternative is to resort to a trigonometric series. Taking into account the boundary conditions (D4.2) and the fact that the maximal deflection takes place at the mid-section of the beam, set

$$w(x) = b_2[1 - \cos(2\pi x/L)] \qquad (D5.2)$$

and repeat the above computation:

```
> d5 := b[2]*(1-cos(2*Pi*x/L)):
> d6 := subs(w(x)=d5,d1):"*d5/b[2]:int(",x=0..L):
> d7 := solve(",b[2]);
```

$$d7 := 2 \frac{qL^4}{3KL^4 + 16 EI\,Pi^4}$$

This solution is thus given by (D5.2) with b_2 specified by the above d7.

To obtain the second approximation modify the expression for w as given below by d9:

```
> d8 := b[4]*(1-cos(4*Pi*x/L)):
```

36 DIRECT METHODS

```
> d9 := d5+d8;
```

$$d9 := b[2]\left(1 - \cos(2\frac{\text{Pi } x}{L})\right) + b[4]\left(1 - \cos(4\frac{\text{Pi } x}{L})\right)$$

There are thus two unknown coefficients, `b[2]` and `b[4]`, to be found from the system of equations involving the weighted residual. These are denoted below as `eq1` and `eq2`:

```
> d10 := subs(w(x)=d9,d1):
> d10*d5/b[2]:eq1 := int(",x=0..L):
> d10*d8/b[4]:eq2 := int(",x=0..L):
```

Solving these equations for the coefficients,

```
> sol := solve({eq1,eq2},{b[2],b[4]});
```

$$sol := \{b[2] = 2\frac{qL^4(256\,EI\,Pi^4 + KL^4)}{4096\,EI^2\,Pi^8 + 816\,EI\,Pi^4\,KL^4 + 5K^2L^8},$$

$$b[4] = 2\frac{qL^4(16\,EI\,Pi^4 + KL^4)}{4096\,EI^2\,Pi^8 + 816\,EI\,Pi^4\,KL^4 + 5K^2L^8}\}$$

one gets the second approximation for the deflection w by substituting the values of the coefficients in `d9`. Note, in the above $\text{EI} = EI$, so $\text{EI}^2 = E^2I^2$.

A general form of approximation for w may be given as

$$w(x) = \sum_{j=1}^{n} b_j[1 - \cos(2j\pi x/L)] \qquad (D5.3)$$

of which the above `d9` is a particular case.

D6. The Rayleigh–Ritz method

(MAPLE file vfem19)

An alternative approach resorts directly to the relevant variational principle rather than to the Euler–Lagrange equation. Indeed, it has been shown in Section D1 that governing equations define the functions, which are extremizers of certain functionals.

Consider the equilibrium of a hinged beam resting on the elastic foundation and subjected to a transverse distributed load q. The potential energy U is

$$U = 1/2 \int_0^L (EI\,w_{,xx}^2 + Kw^2 - 2qw)\,dx \qquad (D6.1)$$

with the notations of Section D4. The first term in (D6.1) is the beam strain energy, the second term is the strain energy of the foundation and the third represents the work done by q. The boundary conditions are

$$w(x=0) = w(x=L) = 0$$
$$w_{,xx}(x=0) = w_{,xx}(x=L) = 0 \qquad (D6.2)$$

Looking for a simple approximate solution, choose a trial function as

$$w = a_1 \sin(\pi x/L) \qquad (D6.3)$$

which obviously satisfies the boundary conditions (D6.2). On substituting (D6.3) in the functional (D6.1), the latter becomes a function of a_1. Consequently, the stationarity condition (in this case the minimum condition) yields

$$dU/da_1 = 0 \qquad (D6.4)$$

Specifying a_1 from this equation and substituting the result back in (D6.3), one arrives at the solution of interest. A better approximation may follow from a general expression

$$w = \sum_{j=1}^{n} a_j \sin(j\pi x/L) \qquad (D6.5)$$

This method is known as the Rayleigh–Ritz method.

Consider again the problem treated in Section D5 and dealing with the clamped beam. A proper approximation is

$$w = (L-x)^2 \sum_{j=2}^{n} a_j x^j \qquad (D6.6)$$

To illustrate better capabilities of MAPLE, write a program which would automatically account for the form of trial function w, the highest power of x attained in (D6.6) n, type of a distributed load q and the beam stiffness EI. To this end, state a procedure, say, `ri`, as given below

```
> ri := proc(w,n,q,EI)
> diff(w,x,x)^2*EI/2+K*w^2/2-q*w:c := int(",x=0..L):
> for j from 2 to n do b[j] := diff(c,a[j]) od:
> end:
```

The first two commands of the procedure formulate the potential energy U according to (D6.1). Then it takes the derivatives of the functional with respect to the coefficients $a_j = $ `a[j]`. These derivatives are denoted as $b_j = $ `b[j]`, $j = 1, 2, \ldots, n$.

Proceed with the computation, specifying the first approximation as

$$w = (L-x)^2 a_2 x^2 \qquad (D6.7)$$

and call for `ri`

38 DIRECT METHODS

```
> ri((L-x)^2*a[2]*x^2,2,q,EI):
```

It remains to solve the equation for a_2 to complete the solution:

```
> d5 := solve(b[2]=0,a[2]):a[2] := simplify(d5);
```

$$a[2] := 21\frac{q}{504EI + KL^4}$$

Note that the result is identical to that of Section D5, namely (D5.1).

Trying a higher approximation, set $n = 4$ and get the third approximation

```
> ri((L-x)^2*sum(a[k]*x^k,k=2..4),4,q,EI):
> solve({b[2]=0,b[3]=0,b[4]=0},{a[2],a[3],a[4]});
```

$$\{a[4] = 1716\frac{KqL^2}{5K^2L^8 + 83952EIKL^4 + 40772160EI^2},$$

$$a[3] = -1716\frac{L^3Kq}{5K^2L^8 + 83952EIKL^4 + 40772160EI^2},$$

$$a[2] = 495\frac{(3432EI + KL^4)q}{5K^2L^8 + 83952EIKL^4 + 40772160EI^2}\}$$

These considerations show that the Bubnov–Galerkin and Rayleigh–Ritz methods are intimately related and constitute two alternative ways of extremizing the same functional. The Bubnov–Galerkin method appears to be more convenient from a technical viewpoint. It applies, being a version of the weighted residual approach, to non-conservative systems too. But it may impose more severe limitations on the type of trial functions than the Rayleigh–Ritz method does. In particular, in the framework of the Rayleigh–Ritz method, the natural boundary conditions may not be taken into account while formulating the trial functions, since one deals with the stationarity of the functional in a direct way.

D7. Master program

(MAPLE file vfem20)

Even though the above programs are capable of treating various cases of the beam bending, including a nonuniformity of the cross-section and external load, they are limited to a given type of the boundary conditions. When these change, one needs another trial function. The use of symbolic manipulation codes enables one to consider the totality of possible cases and write a general subroutine, which treat all of them. A particulate problem may then be solved by simple commands. It seems that the marriage of direct methods and symbolic manipulation codes may essentially modify the former.

Consider the totality of the following cases of beam bending: (1) one end clamped (cantilever), (2) hinged beam, (3) clamped beam, (4) one end clamped, the other hinged. This corresponds to the following boundary conditions:

1) $w(x = 0) = w_{,x}(x = 0) = 0$ \hfill (D7.1)

2) $w(x = 0) = w_{,xx}(x = 0) = 0$ \hfill (D7.2)
$w(x = L) = w_{,xx}(x = L) = 0$

3) $w(x = 0) = w_{,x}(x = 0) = 0$ \hfill (D7.3)
$w(x = L) = w_{,x}(x = L) = 0$

4) $w(x = 0) = w_{,xx}(x = 0) = 0$ \hfill (D7.4)
$w(x = L) = w_{,x}(x = L) = 0$

Except for the cantilever, four boundary conditions are involved. Expanding the deflection w in a Taylor series at the interval $0 \leq x \leq L$, we get

$$w = a_0 + a_1 x + a_2 x^2 + \cdots + a_n x^n \tag{D7.5}$$

If $n > 4$, we may specify some of the coefficients a_j, $j = 0, \ldots, n$, from the above boundary conditions (D7.1)–(D7.4). Futher, since one of the ends must be immovable, say $x = 0$, we may always set $a_0 = 0$, and (D7.5) takes the form

$$w = a_1 x + a_2 x^2 + \cdots + a_n x^n \tag{D7.6}$$

Here $n > 3$, on the understanding that one of the boundary conditions has been taken care of. If, for instance, $n = 5$, then we may still have two coefficients to find from a variational procedure of the above type. Note that the spirit of direct methods requires one to specify the coefficients of the lower powers (frequencies) from the boundary conditions, while leaving the rest for the variational procedure.

In the program below we first state (D7.6), its first and second derivatives, and then the boundary quantities involved in the conditions (D7.1)–(D7.4)

```
> d1 := sum(a[j]*x^{\,j},j=1..5):
> d2 := diff(",x):d3 := diff(",x):
> fL := subs(x=L,d1):f10 := subs(x=0,d2):f1L := subs(x=L,d2):
> f20 := subs(x=0,d3):f2L := subs(x=L,d3):
```

Observe that a boundary value taken on at $x = L$ has been denoted by the last letter L and that taken on at $x = 0$ by 0.

Turning to the variational Rayleigh–Ritz method, specify the `ri`-procedure according to (D6.1)

40 DIRECT METHODS

$$U = 1/2 \int_0^L (EI w_{,xx}^2 + Kw^2 - 2qw)\,dx \tag{D7.7}$$

as follows:

```
> ri := proc(w,q,EI)
> EI*diff(w,x,x)^2/2+K*w^2/2-q*w:int(",x=0..L):
> end:
```

It is seen that the `ri`-procedure allows one to vary its arguments, the trial function w, the load q and the stiffness EI. Its value is the above functional U. Once `ri` has provided the value of U, all that is left to do is to take the derivatives with respect to the free coefficients and solve the system of equations. This approach has the flexibility of the finite element method, which is to be considered in the next chapter. Section D8 deals with relevant examples.

D8. Applications

(MAPLE file vfem20)

As an application of the master program, consider the case of a hinged beam. The relevant boundary conditions are given by (D7.2). We therefore solve these equations for the unknown coefficients a_1, a_2 and a_3 and substitute the result in the expression for w (which is `d1`):

```
> d11 := solve({fL,f20,f2L},{a[1],a[2],a[3]});
```

$$d11 := \{a[2] = 0, a[1] = 7/3\, a[5]\, L^4 + a[4]\, L^3, a[3] = -2/3\, L\, (3\, a[4] + 5\, a[5]\, L)\}$$

```
> w := subs(",d1);
```

$$w := (7/3\, a[5]\, L^4 + a[4]\, L^3)\, x - 2/3\, L\, (3\, a[4] + 5\, a[5]\, L^3)\, x + a[4]\, x^4 + a[5]\, x^5$$

It is seen that a_4 and a_5 have remained 'free' so as to allow for minimization of U. Call for `ri` with the above value of the argument w

```
> d13 := ri(w,q,EI):
```

and then minimize the functional to complete the solution:

```
> d14 := diff(d13,a[4]):d15 := diff(d13,a[5]):
> solve({d14,d15},{a[4],a[5]});
```

$$\{a[5] = 0,\ a[4] = 126\,\frac{q}{31\,KL^4 + 3024\,EI}\}$$

The result for the deflection w is

```
> w := subs(",w);simplify(w);
```

$$126\frac{qx(L^3 - 2L^2x + x^3)}{31KL^4 + 3024EI} \tag{D8.1}$$

The next example is the clamped beam. Referring to the relevant boundary conditions (D7.3), type

```
> solve({fL,f10,f1L},{a[1],a[2],a[3]});
```

$$\{a[1] = 0, a[3] = -L(3a[5]L + 2a[4]), a[2] = 2a[5]L^3 + a[4]L^2\}$$

```
> w := subs(",d1):
```

which shows that a_4 and a_5 are again free. As earlier, call for ri and take the derivatives

```
> d17 := ri(w,q,EI):
> d18 := diff(d17,a[4]):d19 := diff(d17,a[5]):
```

Complete the solution by

```
> solve({d18,d19},{a[4],a[5]}):
> w := subs(",w):simplify(w);
```

$$21\frac{qx^2(L^2 - 2Lx + x^2)}{KL^4 + 504EI} \tag{D8.2}$$

This master program may be further improved. For instance, the differentiation may be performed within the procedure. Here is the modified ri-procedure, the application of which is left to the reader:

```
> ri := proc(w,q,EI)
> EI*diff(w,x,x)^2/2+K*w^2/2-q*w:c := int(",x=0..L):
> for i from 4 to 6 do b[i] := diff(c,a[i]) od:
> end:
```

D9. Considerations of accuracy

The assessment of accuracy is perhaps the most difficult aspect of the direct methods. The conclusion drawn about the accuracy is often of a qualitative nature only. A resort to computerized algebra may improve the situation, as it allows for a relatively easy evaluation of higher-order approximations.

The higher accuracy usually means that the agreement with the exact solution is better in a certain 'average' sense, as comments of Section D show.

Consequently, these solutions may provide much better estimates for various integral quantities, like the total potential energy or the overall stiffness. The error could be more substantial for the derivatives, as they depend on the local behaviour of a function. In particular, a prediction for the deflection should be more accurate than that for the slope, and so forth.

Functionals evaluated for the exact and approximate solutions may obey certain bounds. In the case of a static continuous system, for instance, a 'discrete representation' by a system with a finite number of degrees of freedom is equivalent to the imposition of kinematic constraints. This should render the structure stiffer. Hence, the overall stiffness evaluated via minimization of the potential energy U by the Rayleigh–Ritz method must be larger than (or equal to) the exact stiffness.

A more precise formulation may be given in terms of the work done by the prescribed static forces. In the case, say, of a cantilever subjected to the end force P, this work done on the displacement assumed by the Rayleigh–Ritz method, w_R, is less than (or equal to) the exact work, w_E,

$$Pw_R \leq Pw_E \quad (D9.1)$$

which yields

$$w_R \leq w_E \quad (D9.2)$$

Note, the general rule is that expressed by (D9.1) for this particular case, not (D9.2). This bounding property may not hold for the displacement at an arbitrary point x.

The trial functions must comply with the requirements noted in Sections D4, D5 and D6. In particular, they must satisfy the essential boundary conditions in case of the Rayleigh–Ritz method and the essential and natural boundary conditions in case of the Bubnov–Galerkin method (the version given in Section D6). Thus, the natural boundary conditions, which follow from the variation of the functional, may be omitted when the Rayleigh–Ritz method applies.

A comparison with the exact solution, if possible, may be very instructive. As (D8.2) shows, the result for the clamped beam with $K = 0$ is

$$w = qx^2(L - x)^2/(24EI) \quad (D9.3)$$

which coincides with the exact solution. Let us consider another comparison, for the hinged beam with $KL^4/EI = 4$. The exact deflection at the midsection is $w = 0.012505L^4q/EI$, while (D8.1) provides $w = 0.012508L^4q/EI$. For the bending moment at the same section, the exact value is $M = 0.119914L^2q$, while (D8.1) provides $M = 0.120076L^2q$. Observe that the accuracy of the prediction for deflection is indeed higher than that for its second derivative. Besides the excellent accuracy, note that the exact result for the hinged beam resting on elastic foundation employs hyperbolic functions, in contrast with a simple solution given by (D8.1).

Nevertheless, the accuracy is not always as satisfactory as in these cases; in particular, it may be much less in two- and three-dimensional problems. A further insight into this aspect of direct methods follows from considerations in the next section.

D10. Mathematical considerations

The above methods are mainly heuristic and need an examination from a more rigorous point of view. Another question of interest is the problem inverse to that considered in Section D1, namely, finding a functional for which a given equation is its Euler–Lagrange equation.

Introduce a domain V and its boundary Ω. The domain and its boundary constitute together a close domain V_0. Confining our considerations to functions continuous in V_0, define the so-called *scalar* (or *inner*) *product* of two functions u and v by

$$(u,v) = \int_V u(Q)v(Q)\,dv \qquad (D10.1)$$

where Q is a point of V_0. The properties of the inner product just introduced are similar to those of the scalar product of two vectors. In particular, u and v may be interchanged

$$(u,v) = (v,u) \qquad (D10.2)$$

as follows from (D10.1). Note, the inner product of a function with itself is non-negative and it vanishes if and only if this function is equal to zero. It can be directly verified that if a_1, a_2, b_1 and b_2 are constants, then

$$\begin{aligned}(a_1u_1 + a_2u_2, b_1v_1 + b_2v_2) &= a_1b_1(u_1,v_1) + a_1b_2(u_1,v_2) \\ &+ a_2b_1(u_2,v_1) + a_2b_2(u_2,v_2)\end{aligned} \qquad (D10.3)$$

To further extend the analogy between the 'space' of functions and that of vectors, introduce a quantity analogous to the length of the vector, the *norm* of a function f, by

$$|f| = (f,f)^{1/2} \qquad (D10.4)$$

The norm of a function equals zero if and only if the function is equal to zero. Also,

$$|(u,v)| \leq |u||v| \qquad (D10.5)$$
$$|u+v|^2 \leq |u| + |v| \qquad (D10.6)$$

To prove, say, (D10.6) consider the following computation:

$$|u+v|^2 = (u+v, u+v) = (u,u) + 2(u,v) + (v,v)$$
$$= |u|^2 + |v|^2 + 2(u,v) \leq |u| + |v| \quad \text{(D10.7)}$$

where use has been made of (D10.5), which is known as Schwarz inequality.

These definitions generalize to the case of vector functions U and V, for which the inner product is defined by

$$(U, V) = \int_V (U_x V_x + U_y V_y + U_z V_z)\, dv \quad \text{(D10.8)}$$

and the square of the norm by

$$|U|^2 = (U, U) = \int_V (U_x^2 + U_y^2 + U_z^2)\, dv \quad \text{(D10.9)}$$

D11. Operators

Loosely speaking, an operator is said to be defined if a law is given that specifies a correspondence between each function of this set and one and only one function of another set. These sets of functions merit special names. The set the operator acts upon is the field of its definition, while the resulting set is its field of values. We denote the operators by italic letters.

The operator which has zero field of values is a null operator. The operator I which transforms each element of its definition field to itself is called the identity operator.

Define the sum of operators and their product by

$$(A + B)f = Af + Bf \quad \text{(D11.1)}$$

and

$$ABf = A(Bf) \quad \text{(D11.2)}$$

respectively. Here f is a function from the field of definition of A and B. If

$$ABf = BAf \quad \text{(D11.3)}$$

the operators are permutable. The inverse operator A^{-1} is given by

$$AA^{-1}f = If = f \quad \text{(D11.4)}$$

where I is the identity operator.

Linear operators are of particular interest because of their wide use and simplicity. If, for any f_1 and f_2 in a given set, their sum $(f_1 + f_2)$ and the product (af_1) belong to the same set of functions, then the set is linear. Here a is any scalar. For instance, the set of polynomials and the set of continuous functions are linear. On the other hand, the set of functions for which

$f(x = 10) < 6$ is not linear, as the sum of two such functions may not comply with this property.

An operator A is said to be linear if for any number n and constants $a_1, a_2, \ldots, a_n$ the following expression holds:

$$A \sum_1^n a_i f_i = \sum_1^n A a_i f_i = \sum_1^n a_i A f_i \qquad (D11.5)$$

provided the functions f_i belong to the linear set of the field of definition of A. The identity and null operators are obviously linear.

An operator is symmetric, if for any functions f_1 and f_2 of its field of definition the following relation holds:

$$(Af_1, f_2) = (f_1, Af_2) \qquad (D11.6)$$

provided f_1 and f_2 are sufficiently 'regular' functions and Af_1 and Af_2 have finite norms. For example, the identity operator I is symmetric

$$(If_1, f_2) = (f_1, If_2) = (f_1, f_2) \qquad (D11.7)$$

The null operator is symmetric as well.

As the next example, consider the operator A defined by

$$Af = -f_{,xx} \qquad (D11.8)$$

for functions that together with their derivatives are continuous up to the second order in the segment $0 \leq x \leq 1$. Then

$$(Af_1, f_2) - (f_1, Af_2) = -\int_0^1 (f_2 f_{1,xx} - f_1 f_{2,xx}) \, dx \qquad (D11.9)$$

The integrand is equal to

$$\frac{d}{dx}(f_2 f_{1,x} - f_1 f_{2,x}) \qquad (D11.10)$$

which yields for (D11.9)

$$(Af_1, f_2) - (f_1, Af_2) = -f_2(x=1)f_{1,x}(x=1) + f_1(x=0)f_{2,x}(x=0) \qquad (D11.11)$$

Thus, if the functions f_1 and f_2 vanish at the ends,

$$f_j(x=0) = f_j(x=1) = 0, \quad j = 1, 2 \qquad (D11.12)$$

the above operator A is symmetric.

Finally, a symmetric operator is positive definite if

$$(Af, f) \geq 0 \qquad (D11.13)$$

It is assumed that the above inner product vanishes only if $f = 0$ (some authors use another definition). This type of operator is closely related to the energy of conservative systems.

Consider again the operator defined by (D11.8) and (D11.12). Then

$$(Af,f) = -\int_0^1 f_{,xx} f \, dx \tag{D11.14}$$

This yields, after integration by parts,

$$(Af,f) = \int_0^1 f_{,x}^2 \, dx - f(x=1)f_{,x}(x=1) + f(x=0)f_{,x}(x=0) \tag{D11.15}$$

In view of (D11.12), we get

$$(Af,f) = \int_0^1 f_{,x}^2 \, dx \geq 0 \tag{D11.16}$$

Further, from (D11.12) it follows that the equality in (D11.16) holds for $f = 0$ only. This implies that the operator is positive definite.

The inner product (Af, f) where A is positive definite, may be referred to as the energy of the function f, as similar expressions emerge in the relations describing the energy of conservative systems.

D12. Energy convergence and minimum theorem

The considerations of Section D11 enable one to define a measure of the closeness of the functions, which is suitable for the nature of direct methods. Namely, for the functions f_1 and f_2 from the field of definition of A, define the quantity

$$(Ag, g)^{1/2} \tag{D12.1}$$

where

$$g = f_1 - f_2 \tag{D12.2}$$

Equation (D12.1) is thus the square root of the energy of the difference. This leads to a convergence in energy. The sequence f_n converges in energy to f if

$$\lim_{n \to \infty} (Ag_n, g) \to 0 \tag{D12.3}$$

with

$$g_n = f_n - f \tag{D12.4}$$

In the above A is positive definite and the functions involved must satisfy certain regularity conditions, in particular, they must vanish at the domain

ENERGY CONVERGENCE AND MINIMUM THEOREM

boundary. The energy convergence corresponds to the spirit of direct methods and enables one to establish useful theorems.

Consider an operator equation

$$Af = F \qquad (D12.5)$$

where F is a given function and f an unknown solution, which should also satisfy certain boundary conditions. It is possible to prove that (D12.5) has a unique solution if A is positive definite. Indeed, assume that f_1 and f_2 are solutions to (D12.5):

$$Af_1 = F, \qquad Af_2 = F \qquad (D12.6)$$

Since A is a linear operator, we get, substracting (D21.6),

$$A(f_1 - f_2) = Af_0 = 0 \qquad (D12.7)$$

The energy of $f_0 = f_1 - f_2$ is

$$(Af_0, f_0) = 0 \qquad (D12.8)$$

Since A is a positive definite operator, this implies $f_0 = 0$, $f_1 = f_2$, thereby proving the uniqueness of the solution.

The question arises of which functional would yield (D12.5) as its Euler–Lagrange equation. This problem, which is inverse to that dealt with in Section D1, requires the consideration of functionals, which now appear as particular cases of operators. The definition of a linear functional stems from that of the linear operator given above. The so-called quadratic functional, which is particularly relevant for the subsequent considerations, may be defined by

$$(Af, f) + Lf + C \qquad (D12.9)$$

where A is a linear operator, L is a linear functional and C is a constant. It is seen that (D12.9) 'transforms' f to a constant, in agreement with the definition of a functional given in Section D1.

Consider a quadratic functional associated with (D12.5):

$$W(f) = (Af, f) - 2(f, F) \qquad (D12.10)$$

It is possible to show that as f is varied over the field of definition of A, W takes its least value when f is the solution to (D12.5). To this end, let u be the solution to (D12.5):

$$Au = F \qquad (D12.11)$$

and replace F in (D12.10) by Au:

$$W(f) = (Af, f) - 2(f, Au)) \qquad (D12.12)$$

The direct computation shows that this may be rewritten as

$$W(f) = (A(f-u), f-u) - (Au, u) \qquad (D12.13)$$

with A appearing in both of the terms of the right-hand side. Since A is positive definite, the minimum of (D12.13) occurs when the first term vanishes. This implies $f = u$. The minimum itself is

$$W(f)_{\min} = -(Au, u) \qquad (D12.14)$$

This result is known as the minimum functional theorem. It enables one to search for a minimum of the functional, instead of solving (D12.5). The functional (D12.10) may be proportional to the static potential energy of a system, so the above result does not apply to dynamic phenomena.

To gain further insight into the nature of direct methods, consider a functional $B(u)$ with the exact lower bound C. The sequence u_n is said to be minimizing if

$$\lim_{n \to \infty} B(u_n) = C \qquad (D12.15)$$

It is possible to show that a minimizing sequence for the functional (D12.10) converges in energy to the solution of (D12.5) in a properly defined functional space. Indeed, replacing f by u_n in (D12.13), we get

$$W(u_n) = (A(u_n - u), (u_n - u)) - (Au, u) \qquad (D12.16)$$

which, according to (D12.14) and (D12.15), has the limit

$$\lim_{n \to \infty} W(u_n) = C = -(Au, u) \qquad (D12.17)$$

Therefore,

$$(A(u_n - u), (u_n - u)) \to 0 \qquad (D12.18)$$

which illustrates the convergence in energy, typical of direct methods, and clarifies their mathematical basis. Unfortunately, rigorous statements of this kind are possible for only a relatively narrow class of operators.

D13. More on trial functions

The choice of trial functions is crucial for the success of a direct method and various comments on this subject have been made throughout this chapter. Additional relevant aspects are discussed below.

Two functions f_1 and f_2 are said to be *orthogonal* over V if

$$(f_1, f_2) = 0 \qquad (D13.1)$$

For example, the functions

$$1, \quad \cos x, \quad \sin x, \quad \cos 2x, \quad \sin 2x, \quad \ldots \qquad (D13.2)$$

are orthogonal in the interval $0 \leq x \leq 2\pi$, as may be verified by direct integration. If all functions of an orthogonal set are normalized (have the unit norm), the set is referred to as *orthonormal*. Hence, for the orthonormal set

METHODS OF WEIGHTED RESIDUAL   49

$$(f_n, f_i) = 0, \quad i \neq n$$
$$(f_n, f_i) = 1, \quad i = n.$$
(D13.3)

The functions of an orthonormal set can be shown to be *linearly independent*, which means that the expression

$$a_1 f_1 + a_2 f_2 + \cdots + a_N f_N = 0 \tag{D13.4}$$

holds if and only if all the coefficients a_j vanish.

The notions of orthogonality and linear independence are closely related to direct methods. As (D13.4) shows, if the trial functions are linearly dependent, they may represent a vanishing solution, which might cause computational difficulties. Futher, if the functions are orthogonal, some of the integrals over the domain V may vanish, in agreement with (D13.3). This usually simplifies the calculations in a radical way.

A solution to the equation at hand belongs to a certain class of functions. Therefore, another relevant aspect is the *completeness*, which deals with the question whether a selected set of trial functions is capable of representing any function of the above class. For instance, the set of functions f_k is complete in the sense of convergence in energy if any function with a finite norm can be approximated to in energy within V by a linear combination of f_k.

It should be noted that the cases when an orthonormal and complete set of functions may be used as a trial solution correspond mostly to problems of an academic nature. In the framework of the finite element method, to be treated in Chapter F, these sets may be of use as local approximations or to represent a time dependence. Nevertheless, in the practice of direct methods, trial functions often provide a satisfactory accuracy without being orthonormal and complete.

D14. Methods of weighted residual

The above methods rely either on a relevant functional or on a more formal approach of the weighted residual. The Rayleigh–Ritz method illustrates the first technique and the Bubnov–Galerkin the second. Though a close connection may exist between the two, the technique of weighted residual is presently believed to be more versatile. Below we describe a general scheme of this technique. Assume that the differential equation $Aw = 0$, initial conditions $Nw = 0$ and boundary conditions $Bw = 0$ govern the system. Here A, N and B are linear operators and w an unknown function. On substitution of an approximation w_a for w, one gets the residuals:

$$Aw_a = R_A$$
$$Nw_a = R_N \tag{D14.1}$$
$$Bw_a = R_B$$

50   DIRECT METHODS

Direct methods may be classified as follows: (1) a boundary method if $R_A = 0$ and $R_N = 0$, (2) an interior method if $R_N = 0$ and $R_B = 0$, (3) a mixed method, if only $R_N = 0$. The Rayleigh–Ritz and Bubnov–Galerkin methods belong obviously to the second (interior) type.

In the spirit of direct methods, set for w_a

$$w_a = \sum_{i=0}^{M} a_i(t) f_i(x) \tag{D14.2}$$

where $f_i(x)$ are given spatial functions and $x = x_1, x_2, x_3$. The interior method suggests that the residual R_A must be minimized. The unknown functions (coefficients) $a_i(t)$ may be found from the vanishing inner products

$$(R_A, g_k(x)) = 0, \quad k = 0, 1, \ldots, M \tag{D14.3}$$

where $g_k(x)$ are the so-called *test* functions. The number of equations in (D14.3) corresponds to the number of unknown quantities $a_i(t)$.

Depending on a particular choice of the test functions $g_k(x)$, (D14.3) generates a variety of approximate schemes. If $g_k(x)$ is equal to the derivative of the residual with respect to $a_k(t)$, then (D14.3) effectively amounts to minimization of the inner product

$$(R_A, R_A) = 0, \quad k = 0, 1, \ldots, M \tag{D14.4}$$

This is known as the *least squares* method. If the test functions and the trial functions are from the same set, (D14.3) recovers the classical Bubnov–Galerkin method. A generalized method, when these sets of functions differ from one another is referred to as the Petrov–Galerkin method. Further, setting

$$\begin{aligned} g_k(x) &= 1, & x \text{ in } V_k \\ g_k(x) &= 0, & x \text{ outside } V_k \end{aligned} \tag{D14.5}$$

where V_k is a subdomain of V, one arrives at the *subdomain* method, for which the functions $a_k(t)$ are found so as to make the integral of the residual vanish over each of the subdomains. Instead, one may require the residual R_A to vanish at a selected set of points x_k. This amounts to defining the test functions as

$$g_k(x) = \delta(x - x_k) \tag{D14.6}$$

and is referred to as the *collocation* method.

A boundary method for which $R_A = 0$ and $R_N = 0$ is considered in Section D27.

D15. The iterative Kantorovich method

The Kantorovich method applies to problems of two or more dimensions. Consider a rectangular plate whose edges $x = a$ and $x = -a$ are clamped, and assume that the transverse load is uniform with the intensity q_0. The boundary conditions at the edges $y = b$ and $y = -b$ will be specified later. Represent the unknown deflection $w(x, y)$ as

$$w(x, y) = w_1(x) f(y) \tag{D15.1}$$

where $w_1(x)$ may be specified explicitly as

$$w_1(x) = (x^2 - a^2)^2 \tag{D15.2}$$

By choosing (D15.2) one satisfies the boundary conditions imposed at $x = a$ and $x = -a$. The trial function for $w(x, y)$ becomes

$$w(x, y) = (x^2 - a^2)^2 f(y) \tag{D15.3}$$

In order to find $f(y)$ turn to the equation governing the equilibrium of thin plates

$$\nabla^4 w - q_0/D = 0 \tag{D15.4}$$

with D being the plate stiffness and ∇^4 the biharmonic operator. In agreement with the Bubnov–Galerkin method (or the weighted residual approach), one may set

$$\int_{-a}^{a} \int_{-b}^{b} (\nabla^4 w - q_0/D) \delta w \, dx \, dy = 0 \tag{D15.5}$$

where δw is the variation of $w(x, y)$. According to (D15.3), $\delta w = (x^2 - a^2)^2 \delta f(y)$, and (D15.5) yields

$$\int_{-b}^{b} \left[\int_{-a}^{a} (\nabla^4 w - q_0/D)(x^2 - a^2)^2 \, dx \right] \delta f(y) \, dy = 0 \tag{D15.6}$$

This equation holds for an arbitrary $\delta f(y)$ if the bracketed expression vanishes:

$$\int_{-a}^{a} (\nabla^4 w - q_0/D)(x^2 - a^2)^2 \, dx = 0 \tag{D15.7}$$

Substituting from (D15.3) for w, one gets

$$\int_{-a}^{a} [\nabla^4 (x^2 - a^2)^2 f(y)) - q_0/D](x^2 - a^2)^2 \, dx = 0 \tag{D15.8}$$

which, after integration with respect to x, provides the following differential equation for $f(y)$:

$$a^4 f_{,yyyy} - 6a^2 f_{,yy} - 63f/2 = 21 q_0/(16D) \tag{D15.9}$$

This should be subjected to the boundary conditions at $y = b$ and $y = -b$

relevant for the problem at hand. Note (D15.9) is much simpler than (D15.4). Resolving (D15.9), one obtains the first approximation by the Kantorovich method.

Further, this technique admits iterations. Solving (D15.9) and substituting the solution for $f(y)$ in (D15.1), one may then repeat the computations to find $w_1(x)$. Iterations seem particularly useful for three-dimensional problems for which they may be performed in a variety of ways. The next section contains an example of application of this method to a heat transfer problem.

D16. Heat transfer in a plate

(MAPLE file vfem25 and vfem25c)

Consider a plate, which, under a proper normalization of the coordinate system, occupies the domain $0 \le x \le 1$ and $0 \le y \le \pi$ and radiates heat at the rate $f(x, y)$ per unit area and whose edges are kept at zero temperature. For the sake of certainty, set

$$f(x, y) = b \sin \pi x \sin y \qquad \text{(D16.1)}$$

with b being a constant.

The temperature T obeys the equation

$$T_{,xx} + T_{,yy} = f/k_0 = (b/k_0) \sin \pi x \sin y \qquad \text{(D16.2)}$$

and the homogeneous boundary conditions

$$T(x = 0, y) = T(x = 1, y) = T(x, y = 0) = T(x, y = \pi) = 0 \qquad \text{(D16.3)}$$

Here k_0 is the thermal conductivity. Equation (D16.3) is the Euler–Lagrange equation of the functional

$$F = \int_0^\pi \int_0^1 [T_{,x}^2 + T_{,y}^2 + 2(b/k_0) T \sin \pi x \sin y] \, dx \, dy \qquad \text{(D16.4)}$$

as can be verified by the straightforward computation.

Bearing in mind the application of the Kantorovich method, set the trial function

$$T = x(1 - x) f(y) \qquad \text{(D16.5)}$$

so as to comply with the boundary conditions imposed at $x = 0$ and $x = 1$. The function $f(y)$ should be found from the stationarity of the functional F. In the MAPLE program given below $b/k_0 = 1$. We first state the integrand of F denoted below as u, then substitute for T the expression given by (D16.5) and integrate (using the routine of indefinite integration) with respect to x only, in agreement with the Kantorovich method:

HEAT TRANSFER IN A PLATE

```
> u := "diff(T,x)"^2+"diff(T,y)"^2+2*T*sin(Pi*x)*sin(y);
```

$$u := {'\frac{d}{dx}T'}^2 + {'\frac{d}{dy}T'}^2 + 2T\sin(Pi\,x)\sin(y)$$

```
> c := subs(T=x*(1-x)*f(y),u):
> c;
```

$$((1-x)f(y) - xf(y))^2 + x^2(1-x)^2\left(\frac{d}{dy}f(y)\right)^2$$
$$+ 2x(1-x)f(y)\sin(Pi\,x)\sin(y)$$

```
> int(",x):subs(x=1,")-subs(x=0,"):simplify(");
```

$$1/30\,\frac{10f(y)^2 Pi^3 + \left(\frac{d}{dy}f(y)\right)^2 Pi^3 + 240f(y)\sin(y)}{Pi^3}$$

Denoting the above expression as L, we now get for the functional F

$$F = \int_0^\pi L\,dy \qquad (D16.6)$$

and for its Euler–Lagrange equation (see Section D1)

$$f_{yy} - 10f = (120/\pi^3)\sin y \qquad (D16.7)$$

Note that the derivation of this equation may also be performed in an automated way (see vfem25c.ms). Incorporating the homogeneous boundary conditions with respect to y and solving (D16.7) for $f(y)$, one obtains the first Kantorovich approximation. To this end, state (D16.7) and invoke dsolve:

```
> diff(f(y),y,y)-10*f(y)-120*sin(y)/Pi^3=0;
```

$$\left(\frac{d^2}{dy^2}f(y)\right) - 10f(y) - 120\frac{\sin(y)}{Pi^3} = 0$$

```
> dsolve({",f(0)=0,f(Pi)=0},f(y));
```

$$f(y) = \frac{120}{11}\frac{\sin(y)}{Pi^3} \qquad (D16.8)$$

Substituting this for $f(y)$ in (D16.5) one gets the first approximation.

For the next approximation repeat the computations, this time 'optimizing' the function, which describes the x-dependence. In other words, assume

$$T = q(x)f(y) \qquad (D16.9)$$

with $f(y)$ given by the above (D16.8), and find $q(x)$ so as to extremize the functional F. So, state (D16.9), substitute this in the integrand of F, u, and integrate with respect to y:

```
> s := T=120*q(x)*sin(y)/(11*Pi^3);
```

$$s := T = \frac{120q(x)\sin(y)}{11\,\text{Pi}^3}$$

```
> m := subs(s,u):
> int(",y):subs(y=Pi,")-subs(y=0,"):simplify(");
```

$$\frac{120}{121}\frac{60\left(\dfrac{d}{dx}q(x)\right)^2 + 60q(x)^2 + 11q(x)\sin(\text{Pi}\,x)\,\text{Pi}^3}{\text{Pi}^5}$$

Denoting this expression as N, we now get for the functional F

$$F = \int_0^1 N\,dx \tag{D16.10}$$

and for its Euler–Lagrange equation

$$1440 q_{,xx} - 1440 q = 132\pi^3 \sin \pi x \tag{D16.11}$$

(see vfem25c.ms).

Incorporating the homogeneous boundary conditions, solving (D16.11) and substituting the result for $q(x)$ in (D16.9), complete the solution by

```
> eq := 132*Pi^3*sin(Pi*x)+1440*q(x)-1440*diff(q(x),x,x):
> dsolve({eq,q(0)=0,q(1)=0},q(x));
```

$$q(x) = -\frac{11}{120}\frac{\text{Pi}^3\sin(\text{Pi}\,x)}{1+\text{Pi}^2}$$

```
> Temperature(x,y) := subs(",s);
```

$$\text{Temperature}(x,y) := T = -\frac{\sin(\text{Pi}\,x)\sin(y)}{1+\text{Pi}^2}$$

This coincides with the exact solution. Of course, in case of three or more dimensions the iterations may not be limited to the second order and may bring about particularly improved results.

D17. Constraints

The presence of various constraints is typical of engineering problems. Besides the boundary conditions, which may also be considered as constraints, these might include the volume (weight) of the structure, its maximal response, kinematic relations and others. Problems of this kind may be treated with the help of *Lagrange multipliers*.

Consider first a simple example of the wheel rolling down on a plane (Fig. D3). If no slipping takes place, the kinetic energy T is

$$T = mx_{,t}^2/2 + ma^2\xi_{,t}^2/2 \tag{D17.1}$$

where m is the wheel mass and a is the radius. The potential energy U is

$$U = mg(L - x)\sin\phi \tag{D17.2}$$

(see Fig. D3 for the remaining notations). Because of pure rolling, the coordinates x and ξ are related by

$$a\xi - x = 0 \tag{D17.3}$$

which is a kinematic constraint.

State the modified Hamilton's functional, which also includes (D17.3),

$$I = \int_{t_1}^{t_2} [T - U + \lambda(a\xi - x)]\,dt \tag{D17.4}$$

with λ being the unknown yet Lagrange multiplier. Since (D17.3) vanishes and λ is assumed to be finite, the value of the functional has not been changed. Considering x and ξ as the *independent coordinates*, write the Euler–Lagrange equations (D1.10)

$$ma\xi_{,tt} - \lambda = 0$$
$$mx_{,tt} - mg\sin\phi + \lambda = 0 \tag{D17.5}$$

Equations (D17.3) and (D17.5) provide $\lambda = mx_{,tt}$ and

$$x_{,tt} = g(\sin\phi)/2, \qquad \xi_{,tt} = g(\sin\phi)/(2a) \tag{D17.6}$$

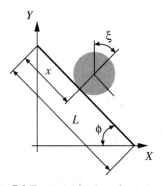

D3 To constrained motions

which comply with (D17.3) and, after integration, yield the desirable coordinates $x = x(t)$ and $\xi = \xi(t)$.

It is seen that this technique manages to extremize the functional and, at the same time, to comply with the constraint (D17.3). It enables one to treat the problem as *unconstrained* by introducing the modified functional (D17.4), but leads to a larger number of unknown quantities. Indeed, with a new condition one must introduce a new Lagrange multiplier. In general, (D17.4) acquires the form

$$I = \int_{t_1}^{t_2} [T - U + \lambda_i s_i(x_1, x_2, \ldots)] \, dt \qquad (D17.7)$$

where $s_i(x_1, x_2, \ldots) = 0$ are the constraints and the repeated index i implies summation. In physical terms, the Lagrange multipliers λ_i express the generalized forces associated with the constraints. Note, the constraints

$$s_i(x_1, x_2, \ldots) = 0 \qquad (D17.8)$$

also follow from the stationarity condition $\partial I / \partial \lambda_i = 0$.

A constraint may also have the form of inequality, for example,

$$y_{,x} \leq M \qquad (D17.9)$$

Introducing a real variable a, set instead of (D17.9)

$$y_{,x} + a^2 - M = 0 \qquad (D17.10)$$

which may be accounted for by the method presented.

D18. Computer-generated Euler–Lagrange equations

(MAPLE file vfem9)

Since the Euler–Lagrange equations follow in a unique way once a generic functional is given, a computer may take over their derivation. Below we begin with a program which derives a single equation, and then extend this approach to systems of two degrees of freedom.

Starting with the harmonic oscillator, formulate its Lagrangian, la, and then its derivatives

```
> la := m*ht^2/2-k*h^2/2;
```

$$la := 1/2 \, m \, ht^2 - 1/2 k h^2$$

```
> q := diff(la,ht); f := diff(la,h);
  s := subs({ht=diff(h(t),t),h=h(t)},q);
```

COMPUTER-GENERATED EULER–LAGRANGE EQUATIONS

```
q := m h
f := -k h
s := m ( d/dt h(t) )
```

Here h is the generalized coordinate (h = x) and ht the velocity (ht = $x_{,t}$). Recalling and adjusting (D1.10), we get the Euler–Lagrange equation by

```
> c := subs({ht=diff(h(t),t),h=h(t)},f);
  c := -k h(t)

> E_L := c-diff(s,t);
  E_L := -k h(t) - m ( d²/dt² h(t) )
```

This program applies to other systems of a single degree of freedom, provided the Lagrangian is properly specified. For the nonlinear pendulum the kinetic energy is

$$T = mL^2 \theta_{,t}^2 / 2 \tag{D18.1}$$

and the potential energy is

$$U = mgL(1 - \cos\theta) \tag{D18.2}$$

where θ is the rotation angle. Replacing θ by h and $\theta_{,t}$ by ht, state the Lagrangian as

```
> la := m*L^2*ht^2/2-m*g*L*(1-cos(h));
  la := 1/2 m L² ht² - m g L(1 - cos(h))
```

which may be done with the help of the Editor, and run the program once again:

```
> q := diff(la,ht);
  s := subs({ht=diff(h(t),t),h=h(t)},q);
  q := m L² ht
  s := m L² ( d/dt h(t) )

> c := subs({ht = diff(h(t),t), h = h(t)}, diff(la,h));
  c := - m g L sin(h(t))

> E_L := c-diff(s,t) : E_L := simplify(E_L/(m*L));
  E_L := -g sin(h(t)) - L ( d²/dt² h(t) )
```

58 DIRECT METHODS

which is the right-hand side of the governing equation.

Furthermore, the program can be modified so as to allow for an automatic derivation with the help of `proc`:

```
> w := proc(la)
> q := diff(la,ht); f := diff(la,h); o := diff(la,htt);
> d := {htt=diff(h(t),t,t),ht=diff(h(t),t),h=h(t)};
> c := subs(d,f); s := subs(d,q); r := subs(d,o);
> E_L := c-diff(s,t)+diff(r,t,t);
> end;
```

For example, we get the same result for the above pendulum:

```
> w(la);
```

$$-m\,g\,L\sin(h(t)) - m\,L^2\left(\frac{d^2}{dt^2}h(t)\right)$$

D19. Two degrees of freedom

(MAPLE file vfem10)

The previous program modifies easily to the case of systems of two degrees of freedom. Consider a double pendulum as shown in Fig. D4.

In the notation of this figure the potential energy is

$$U = -m_2 g L_2 \cos\theta_2 - (m_1 + m_2) g L_1 \cos\theta_1 \tag{D19.1}$$

and the kinetic energy is

$$T = (m_1 + m_2) L_1^2 \theta_{1,t}^2 / 2 + m_2 L_2^2 \theta_{2,t}^2 / 2 \\ + m_2 L_1 L_2 \theta_{1,t} \theta_{2,t} \cos(\theta_1 - \theta_2) \tag{D19.2}$$

Denoting θ_1 by h1, θ_2 by h2, $\theta_{1,t}$ by h1t and $\theta_{2,t}$ by h2t for the coordinates and velocities, specify the Lagrangian, la, and then other relevant relations,

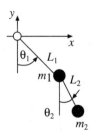

D4 Double pendulum

CONTINUOUS SYSTEMS   59

similar to those of the previous section. In anticipation of lengthy expressions, avoid the display of these intermediate results:

```
> la := (m1+m2)*L1^2*h1t^2/2+m2*L2^2*h2t^2/2
    +m2*L1*L2*h1t*h2t*cos(h1-h2)+
> m2*g*L2*cos(h2)+(m1+m2)*g*L1*cos(h1):
> q1 := diff(la,h1t):q2 := diff(la,h2t):f1 := diff(la,h1):
    f2 := diff(la,h2):
> d := {h1t=diff(h1(t),t),h2t=diff(h2(t),t),h1=h1(t),h2=h2(t)}:
> s1 := subs(d,q1):c1 := subs(d,f1):s2 := subs(d,q2):
    c2 := subs(d,f2):
```

The Euler–Lagrange equations are

```
> E_L1 := c1-diff(s1,t):
> E_L2 := c2-diff(s2,t):
```

which, after simplification, yield

```
> E_L1 := simplify(E_L1);
    E_L1 := -g*L1*sin(h1(t))*m1 - g*L1*sin(h1(t))*m2
    - L1**2*diff(diff(h1(t),t),t)*m1
    - L1**2*diff(diff(h1(t),t),t)*m2
    - m2*L1*L2*diff(diff(h2(t),t),t)*cos(h1(t)-h2(t))
    - m2*L1*L2*diff(h2(t),t)**2*sin(h1(t) - h2(t))
> E_L2 := simplify(E_L2);
    E_L2 := - m2*g*L2*sin(h2(t))-m2*L2**2*diff(diff(h2(t),t),t)
    - m2*L1*L2*diff(diff(h1(t),t),t)*cos(h1(t)- h2(t))
    + m2*L1*L2*diff(h1(t),t)**2*sin(h1(t)- h2(t))
```

By changing the expression for the Lagrangian with the help of the Editor, one may investigate other systems of two degrees of freedom.

D20. Continuous systems

(*MAPLE file vfem12*)

Hamilton's principle, when applied to continuous systems, yields (D2.15), reproduced immediately below (summation with respect to $j = 1, 2, 3$ applies):

$$d(\partial L_0/\partial \eta_{,t})/dt + d(\partial L_0/\partial \eta_{,j})/dx_j - \partial L_0/\partial \eta = 0 \qquad (D2.15)$$

If the Lagrangian density L_0 depends also on the second spatial derivatives, (D2.15) properly modifies to

$$-d^2(\partial L_0/\partial \eta_{,jj})/dx_j^2 + d(\partial L_0/\partial \eta_{,t})/dt + d(\partial L_0/\partial \eta_{,j})/dx_j$$
$$- \partial L_0/\partial \eta = 0 \qquad (D20.1)$$

60   DIRECT METHODS

It is seen that these equations may be quite complicated and a computerized derivation is therefore particularly desirable.

Besides the unknown function $\eta(x_1, x_2, x_3, t)$ and its spatial and temporal derivatives, the Lagrangian L_0 may depend on quantities like the density, ρ, moment of inertia I, cross-sectional area S and so forth. These, in turn, may be spatially dependent quantities.

Consider a thin elastic plate under the distributed lateral load q, for which the potential energy density is

$$U_0 = D/2[(w_{,xx} + w_{,yy})^2 - 2(1-v)(w_{,xx}w_{,yy} - w_{,xy}^2)] - qw \qquad (D20.2)$$

Here w is the deflection, v is Poisson's ratio and D is

$$D = EH^3/[12(1-v^2)] \qquad (D20.3)$$

where H is the plate thickness and E Young's modulus. The kinetic energy density T_0 is

$$T_0 = \rho w_{,t}^2/2 \qquad (D20.4)$$

where ρ is the mass density per unit area. Equations (D20.2) and (D20.4) specify the Lagrangian density $L_0 = T_0 - U_0$.

Going over to symbolic computations, state $L_0 = $ L0 as

```
> L0 := h*q(x,y,t)+(-hxy^2+hxx*hyy)*(1-m)*sd-
   (hxx+hyy)^2*sd/2+ro*ht^2/2:
```

with $h = w$, $hxx = w_{,xx}$, $hyy = w_{,yy}$ and $sd = D$ and compute, in a way similar to that of Section D19, the other relevant quantities appearing in (D20.1). For the sake of brevity, we present below the input commands only:

```
> Lh := diff(L0,h);Lht := diff(L0,ht);
   Lhx := diff(L0,hx);Lhy := diff(L0,hy);
> Lhxx := diff(L0,hxx);Lhyy := diff(L0,hyy);
   Lhxy := diff(L0,hxy);
> f := h(x,y,t);
> d := {h=h(x,y,t),ht=diff(f,t),hx=diff(f,x),hy=diff(f,y),
   hxx=diff(f,x,x),
> hyy=diff(f,y,y),hxy=diff(f,x,y)};
> qh := subs(d,Lh);qht := subs(d,Lht);qhx := subs(d,Lhx);
   qhy := subs(d,Lhy);
> qhxx := subs(d,Lhxx);qhyy := subs(d,Lhyy);
   qhxy := subs(d,Lhxy);
```

We get the Euler–Lagrange equation by

```
> E_L := diff(qht,t)-diff(qhxx,x,x)-diff(qhyy,y,y)-
   diff(qhxy,x,y)+diff(qhx,x)+diff(qhy,y)-qh;
```

$$E\_L := ro\ D[3, 3](h)(x, y, t) + (D[1, 1, 1, 1](h)(x, y, t)$$
$$+ D[1, 1, 2, 2](h)(x, y, t))\ sd + (D[1, 1, 2, 2](h)(x, y, t)$$
$$+ D[2, 2, 2, 2](h)(x, y, t))\ sd - q(x, y, t)$$

where $D[i](h)(x, y, t)$ denotes the derivative of $h(x, y, t)$ with respect to its ith argument. This simplifies to

> simplify(");

$$ro\ D[3, 3](h)(x, y, t) + sd\ D[1, 1, 1, 1](h)(x, y, t) + 2\ sd\ D[1, 1, 2, 2](h)(x, y, t)$$
$$+ sd\ D[2, 2, 2, 2](h)(x, y, t) - q(x, y, t)$$

> E_L := collect(",sd);

$$E\_L := (D[1, 1, 1, 1](h)(x, y, t) + 2\ D[1, 1, 2, 2](h)(x, y, t)$$
$$+ D[2, 2, 2, 2](h)(x, y, t))\ sd + ro\ D[3, 3](h)(x, y, t) - q(x, y, t)$$

which is the well-known equation of vibrations of a thin elastic plate

$$D\nabla^4 w(x, y, t) + \rho w_{,tt}(x, y, t) = q(x, y, t) \qquad (D20.5)$$

with $D = $ sd given by (D20.3).

D21. Automatic derivation of governing equations

(MAPLE files vfem13 and vfem14))

The proc may help to further simplify the automatic generation of the governing equations. Below we present a program which modifies that of Section D19 and provides the Euler–Lagrange equations for a system of two degrees of freedom in an automatic way, if the Lagrangian is given:

```
> w := proc(la)
> q1 := diff(la,h1t);q2 := diff(la,h2t);
    f1 := diff(la,h1);f2 := diff(la,h2);
> d := {h1t=diff(h1(t),t),h2t=diff(h2(t),t),
    h1=h1(t),h2=h2(t)};
> s1 := subs(d,q1);c1 := subs(d,f1);s2 := subs(d,q2);
    c2 := subs(d,f2);
> E_L1 := c1-diff(s1,t);
> E_L2 := c2-diff(s2,t);
> E_L1 := simplify(E_L1);
> E_L2 := simplify(E_L2);
> print(E_L1,E_L2);
> end:
```

Verifying this program, specify the Lagrangian la of the double pendulum, as given by (D19.1) and (D19.2), and call for the above *w*-procedure

62 DIRECT METHODS

```
> la := (m1+m2)*L1^2*h1t^2/2+m2*L2^2*h2t^2/2
    +m2*L1*L2*h1t*h2t*cos(h1-h2)+
> m2*g*L2*cos(h2)+(m1+m2)*g*L1*cos(h1):
> w(la);
    -g*L1*sin(h1(t))*m1 - g*L1*sin(h1(t))*m2
    -L1**2*diff(diff(h1(t),t),t)*m1
    -L1**2*diff(diff(h1(t),t),t)*m2
    -m2*L1*L2*diff(diff(h2(t),t),t)*cos(h1(t) - h2(t))
    -m2*L1*L2*diff(h2(t),t)**2*sin(h1(t) - h2(t)),
    -m2*g*L2*sin(h2(t)) - m2*L2**2*diff(diff(h2(t),t),t)
    -m2*L1*L2*diff(diff(h1(t),t),t)*cos(h1(t)
    -h2(t))+m2*L1*L2*diff(h1(t),t)**2*sin(h1(t) - h2(t))
```

Similarly, modify the program of Section 20 for continuous systems:

```
> w := proc(L0)
> Lh := diff(L0,h); Lht := diff(L0,ht);
    Lhx := diff(L0,hx);Lhy := diff(L0,hy);
> Lhxx := diff(L0,hxx);Lhyy := diff(L0,hyy);
    Lhxy := diff(L0,hxy);
> f := h(x,y,t);
> d := {h=h(x,y,t),ht=diff(f,t),hx=diff(f,x),
    hy=diff(f,y),hxx=diff(f,x,x),
> hyy=diff(f,y,y),hxy=diff(f,x,y)};
> qh := subs(d,Lh);qht := subs(d,Lht);qhx := subs(d,Lhx);
    qhy := subs(d,Lhy);
> qhxx := subs(d,Lhxx);qhyy := subs(d,Lhyy);
    qhxy := subs(d,Lhxy);
> E_L := diff(qht,t)-diff(qhxx,x,x)-diff(qhyy,y,y)-
    diff(qhxy,x,y)+
> diff(qhx,x)+diff(qhy,y)-qh;
> collect(",m):collect(",sd):
> print(");
> end;
```

Verifying this program, specify the Lagrangian density L0 of the thin elastic plate, as given by (D20.2) and (D20.4), and call for the above w-procedure:

```
> L0 := h*q(x,y,t)+(-hxy^2+hxx*hyy)*(1-m)*sd -
    (hxx+hyy)^2*sd/2+ro*ht^2/2;
L0 := h*q(x,y,t)+(-hxy**2 + hxx*hyy)*(1-m)*sd
    - 1/2*(hxx + hyy)**2*sd + 1/2*ro*ht**2
> w(L0);
    (((D[1,1,1,1])(h))(x,y,t) + 2*((D[1,1,2,2])(h))(x,y,t)
    + ((D[2,2,2,2])(h))(x,y,t))*sd +
    ro*((D[3,3])(h))(x,y,t) - q(x,y,t)
```

where, as earlier, $D[i](h)(x, y, t)$ denotes the derivative of $h(x, y, t)$ with respect to its ith argument. In both cases the results are identical to those previously derived.

D22. Nature of extremum

Since by changing the sign of the functional one transforms a minimum to a maximum and vice versa, it is sufficient to consider, say, a minimum only. Denote a minimum of the functional J given by

$$J = \int_{x_1}^{x_2} f(x, y, y_{,x}) \, dx \tag{D22.1}$$

as J^0 and assume that $y^0(x)$ is the extremizer. The positive quantity ΔJ given by

$$\Delta J = J - J^0 \tag{D22.2}$$

is the increment of the functional J, which may be set as

$$\Delta J = \int_{x_1}^{x_2} \Delta f(x, y, y_{,x}) \, dx \tag{D22.3}$$

with

$$\Delta f(x, y, y_{,x}) = f(x, y, y_{,x}) - f(x, y^0, y^0_{,x}) \tag{D22.4}$$

This increment stems from the variations $\delta y(x)$ and $\delta y_{,x}(x)$, which, according to (D1.4), are

$$\begin{aligned} \delta y(x) &= y(x) - y^0(x) = \epsilon q(x) \\ \delta y_{,x}(x) &= y_{,x}(x) - y^0_{,x}(x) = \epsilon q_{,x}(x) \end{aligned} \tag{D22.5}$$

where $y(x)$ is an admissible function and ϵ may be thought of as a small parameter. Assuming that $\delta y(x)$ and $\delta y_{,x}(x)$ are also small (the case of weak variations), represent $\Delta f(x, y, y_{,x})$ with the help of a Taylor expansion:

$$\Delta f = \delta f + \delta^2 f / 2 + \cdots \tag{D22.6}$$

with

$$\begin{aligned} \delta f &= f_{,y} \, \delta y + f_{,y'} dy' \\ \delta^2 f &= f_{,yy} (\delta y)^2 + 2 f_{,yy'} \delta y \delta y' + f_{,y'y'} (\delta y')^2 \end{aligned} \tag{D22.7}$$

Here $y' = y_{,x}$ etc. The quantities δf and $\delta^2 f$ are the first and second variations of f, respectively.

Now (D22.3) takes a form similar to (D22.6):

$$\Delta J = \delta J + \delta^2 J / 2 + \cdots \tag{D22.8}$$

64 DIRECT METHODS

with

$$\delta J = \int_{x_1}^{x_2} \delta f(x, y, y_{,x}) \, \mathrm{d}x \tag{D22.9}$$

$$\delta^2 J = \int_{x_1}^{x_2} \delta^2 f(x, y, y_{,x}) \, \mathrm{d}x \tag{D22.10}$$

being the first and second variations of the functional J.

On the other hand, according to (D1.11) the first variation is

$$\delta J = \frac{\mathrm{d}J}{\mathrm{d}\epsilon}(\epsilon = 0)\epsilon \tag{D22.11}$$

which suggests the following expression for the second variation:

$$\delta^2 J = \frac{\mathrm{d}^2 J}{\mathrm{d}\epsilon^2}(\epsilon = 0)\epsilon^2 \tag{D22.12}$$

Consequently, (D22.8) represents a Taylor series about $\epsilon = 0$.

Assume for the moment that $\delta J \neq 0$ for $J = J^0$, which is a stationary value, say, a minimum. Then δJ changes sign, if $\delta y(x)$ does. Indeed, (D22.7) and (D22.9) show that δy and $\delta y'$ govern the value of δJ. But

$$\delta y' = \mathrm{d}\epsilon q(x)/\mathrm{d}x = \mathrm{d}\delta y(x)/\mathrm{d}x \tag{D22.13}$$

where use has been made of (D22.5). Consequently, contrary to the basic assumption, J^0 is not a minimum of J, which implies $\delta J = 0$ as the condition of stationarity. This is in agreement with the conclusion made in Section D1. It is seen from (D22.8) that the second variation $\delta^2 J$ governs the behaviour of J in the vicinity of J^0. For $\delta^2 J > 0$ the stationary value J^0 is a minimum and $\delta^2 J < 0$ a maximum.

D23. Transversality conditions

The natural boundary conditions introduced in Section D3 follow from a variational procedure, they are not imposed *a priori*. The more general transversality conditions also follow from a variational process.

To this end, remove the limitations of the so-called fixed endpoint problems considered so far, namely, assume that the initial point $(x_1, y_1 = y(x_1))$ and the final point $(x_2, y_2 = y(x_2))$ are not specified from the beginning but belong to arbitrary (sufficiently smooth) curves $f_1(x, y)$ and $f_2(x, y)$ shown in Fig. D5. This is referred to as the variable endpoint problem. It is seen that the variation of the admissible functions, Δy, does not necessarily vanish at x_1 and x_2, unlike the fixed endpoint problem.

TRANSVERSALITY CONDITIONS

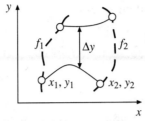

D5 To transversality conditions

First, derive the relation between Δy given by

$$\Delta y = y(x+\delta x) - y^0(x) \tag{D23.1}$$

and the previously specified variation δy given by (D22.5):

$$\delta y(x) = y(x) - y^0(x) = \epsilon q(x) \tag{D22.5}$$

From the above relations, we get

$$\Delta y = y(x+\delta x) - y(x) + \delta y(x) \tag{D23.2}$$

Since

$$y(x+\delta x) - y(x) = y_{,x}(x)\,\delta x + \cdots \tag{D23.3}$$

(D23.2) yields the desirable relation

$$\delta y(x) = \Delta y - y_{,x}(x)\,\delta x \tag{D23.4}$$

Applying the rule of differentiation of an integral with variable limits to (D22.1)

$$\delta J = \int_{x_1}^{x_2} \delta f(x,y,y_{,x})\,dx + [f(x,y,y_{,x})\delta x]_{x=x_2} - [f(x,y,y_{,x})\delta x]_{x=x_1} \tag{D23.5}$$

where the last two terms follow from the variation of the endpoints. The first term on the right-hand side may be set as

$$\int_{x_1}^{x_2} [\partial f/\partial y - d(\partial f/\partial y_{,x})/dx]\delta y(x)\,dx + (\partial f/\partial y_{,x})\delta y(x)_{x=x_2}$$
$$- (\partial f/\partial y_{,x})\delta y(x)_{x=x_1} \tag{D23.6}$$

where results of Section D1, in particular (D1.9), have been used. In view of (D23.4), this yields

$$\int_{x_1}^{x_2} [\partial f/\partial y - d(\partial f/\partial y_{,x})/dx]\delta y(x)\,dx + (\partial f/\partial y_{,x})[\Delta y - y_{,x}(x)\delta x]_{x=x_2}$$
$$- (\partial f/\partial y_{,x})[\Delta y - y_{,x}(x)\delta x]_{x=x_1} \tag{D23.7}$$

Equation (D23.5) eventually becomes

$$\delta J = \int_{x_1}^{x_2} [\partial f/\partial y - \mathrm{d}(\partial f/\partial y_{,x})/\mathrm{d}x]\delta y \, \mathrm{d}x + T \tag{D23.8}$$

where T is

$$T = (\partial f/\partial y_{,x})_{x=x_2}\Delta y_2 - (\partial f/\partial y_{,x})_{x=x_1}\Delta y_1 \\ + [f - (\partial f/\partial y_{,x})y_{,x}]_{x=x_2}\delta x_2 - [f - (\partial f/\partial y_{,x})y_{,x}]_{x=x_1}\delta x_1 \tag{D23.9}$$

The first term on the right-hand side of (D23.8) represents the Euler–Lagrange equation, which holds for the variable endpoint problem too. The remaining term, T, which contains four boundary variations, namely, δx_1, δx_2, Δy_1 and Δy_2, constitute the *transversality* conditions. The correct statement of the problem should not therefore include more than four boundary conditions. If all of these are given, the above variations and, consequently, the transversality conditions, vanish.

D24. Generalizations

The functional to be extremized may not have the form given by (D1.2). Consider the case when the quantity of interest is a ratio of two functionals:

$$J_1/J_2 = \int_{x_1}^{x_2} f_1(x, y, y_{,x}) \, \mathrm{d}x \Big/ \int_{x_1}^{x_2} f_2(x, y, y_{,x}) \, \mathrm{d}x \tag{D24.1}$$

Such a case arises, in particular, in vibration problems. The first variation is

$$\delta(J_1/J_2) = (J_2\delta J_1 - J_1\delta J_2)/J_2^2 \tag{D24.2}$$

Denoting the extremizer as $y^0(x)$, we get, for (D24.1),

$$g = J_1(y^0)/J_2(y^0) \tag{D24.3}$$

which yields for the numerator of (D24.2)

$$\delta J_1 - g\delta J_2 = \delta \int_{x_2}^{x_1} (f_1 - gf_2) \, \mathrm{d}x = 0 \tag{D24.4}$$

The Euler–Lagrange equation associated with (D24.4) is

$$f_{1,y} - gf_{2,y} - \mathrm{d}(f_{1,y'} - gf_{2,y'})/\mathrm{d}x \tag{D24.5}$$

where, y' denotes the derivative with respect to $y_{,x}$.

Set, for example,

$$J_1 = \int_0^1 y_{,x}^2 \, dx, \quad J_2 = \int_0^1 y^2 \, dx \quad (D24.6)$$

with the boundary conditions

$$y(0) = y(1) = 0 \quad (D24.7)$$

Since $f_1 = y_{,x}^2$ and $f_2 = y^2$, (D24.5) gives

$$y_{,xx} + gy = 0 \quad (D24.8)$$

with a solution

$$y(x) = \sin(g^{1/2}x) \quad (D24.9)$$

and $g = \pi^2, 4\pi^2, 9\pi^2, \ldots, n^2\pi^2, n = 1, 2, 3, \ldots$.

Another quantity of interest is

$$J(y) = \int_{x_1}^{x_2} f(x, y, y_{,x}) \, dx + F(x_1, y_1, x_2, y_2) \quad (D24.10)$$

where F is a given function. Though this is different from (D1.2), the Euler–Lagrange equation remains obviously the same. On the other hand, the following term:

$$(\partial F/\partial x_1)\delta x_1 + (\partial F/\partial x_2)\delta x_2 + (\partial F/\partial y_1)\Delta y_1 + (\partial F/\partial y_2)\Delta y_2 \quad (F24.11)$$

must be added to the transversality condition given by (D23.9).

D25. Gauss' principle

Hamilton's principle considered in Section D2 belongs, generally speaking, to the class of stationarity principles, in the sense that the exact nature of the extremum may not be known. Unlike that, Gauss' principle deals with a functional, which yields the governing equations as a condition of its minimum.

To begin, consider a system of particles and define the so-called Gauss' variation as follows:

$$\delta \mathbf{r}_i = \delta \mathbf{r}_{i,t} = \delta \mathbf{F}_i = 0$$
$$\delta \mathbf{r}_{i,tt} \neq 0 \quad (D25.1)$$

where $\mathbf{r}_i$ and $\mathbf{F}_i$ are the generalized coordinates and forces, respectively. In other words, under this particular variation, the admissible motions differ from each other by acceleration only, while the position, velocity and force are identical.

D'Alambert's principle yields

$$\sum_i (\mathbf{F}_i - m_i \mathbf{a}_i)\, \delta \mathbf{r}_i = 0 \qquad (D25.2)$$

where $\delta \mathbf{r}_i$ is a virtual displacement consistent with the imposed constraints. Expanding $\mathbf{r}_i(t)$, we get for the instant $(t + t_1)$

$$\mathbf{r}_i(t + t_1) = \mathbf{r}_i(t) + \mathbf{r}_{i,t}(t) t_1 + \mathbf{a}_i(t) t_1^2/2 + \cdots \qquad (D25.3)$$

In view of (D25.1), this gives

$$\delta \mathbf{r}_i(t + t_1) = \delta \mathbf{a}_i(t) t_1^2/2 \qquad (D25.4)$$

and (D25.2) takes the form

$$\sum_i (\mathbf{F}_i - m_i \mathbf{a}_i) \delta \mathbf{a}_i = 0 \qquad (D25.5)$$

Since $\delta \mathbf{F}_i = 0$ (D25.1), this relation may be rewritten as

$$\sum_i (\mathbf{F}_i - m_i \mathbf{a}_i) \delta (\mathbf{F}_i/m_i - \mathbf{a}_i) = 0 \qquad (D25.6)$$

which in turn takes the form

$$\delta \sum_i (\mathbf{F}_i - m_i \mathbf{a}_i)^2/(2 m_i) = 0 \qquad (D25.7)$$

This stationarity condition is in fact a condition of minimum, as the variation applies to the sum of positive terms. Denoting

$$Z = \sum_i (\mathbf{F}_i - m_i \mathbf{a}_i)^2/(2 m_i) \qquad (D25.8)$$

one may state that the actual motion is such that the 'constraint' Z becomes as small as possible. If Z vanishes (the particles are free to move), one recovers Newton's law:

$$\mathbf{F}_i = m_i \mathbf{a}_i \qquad (D25.9)$$

This interpretation shows that Gauss' principle may be thought of as the least squares method (see Section D14) applied to Newton's law.

D26. Minimum pressure drag

(MAPLE file vfem15)

Fig. D6 shows a body of revolution in a gas flow. The pressure on its surface is

$$p = 2\rho v^2 \sin^2 \phi \qquad (D26.1)$$

MINIMUM PRESSURE DRAG   69

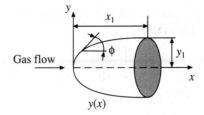

D6 Body of revolution in a gas flow

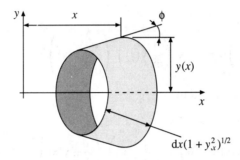

D7 Elementary ribbon

where the angle ϕ is shown in the fiqure, v is the relative speed and ρ the fluid density. Considerations of the elementary ribbon of the width $(1+y_{,x}^2)^{1/2}\,dx$ and radius $y(x)$ (Fig. D7) lead to the following expression for the elementary force:

$$dF = p2\pi y(1+y_{,x}^2)^{1/2} \sin\phi\, dx \qquad (D26.2)$$

which, in view of (D26.1), yields the total force

$$F = 4\pi\rho v^2 \int_0^{x_1} y(1+y_{,x}^2)^{1/2} \sin^3\phi\, dx \qquad (D26.3)$$

where x_1 is the body length. For a slender body

$$\sin\phi = y_{,x}/(1+y_{,x}^2)^{1/2} \approx y_{,x} < 1 \qquad (D26.4)$$

This provides for F

$$F = 4\pi\rho v^2 \int_0^{x_1} y_{,x}^3 y\, dx \qquad (D26.5)$$

when $y(x)$ satisfies the boundary conditions

$$y(x=0) = 0, \qquad y(x=x_1) = y_1 \qquad (D26.6)$$

70 DIRECT METHODS

In order to find $y(x)$, which would provide a minimum to (D26.5), write its Euler–Lagrange equation (D1.10):

$$y_{,x}^3 - d(3y_{,x}^2 y)/dx = 0 \qquad (D26.7)$$

This is a non-linear equation, which simplifies by the following explicit evaluation of the last term:

> y(x)*diff(y(x),x)^2;

$$y(x)\left(\frac{d}{dx}y(x)\right)^2$$

> diff(",x);

$$\left(\frac{d}{dx}y(x)\right)^3 + 2\,y(x)\left(\frac{d}{dx}y(x)\right)\left(\frac{d^2}{dx^2}y(x)\right)$$

Thus (D26.7) takes the form

> diff(y(x),x)^3-3*";

$$\left(-2\frac{d}{dx}y(x)\right)^3 - 6\,y(x)\left(\frac{d}{dx}y(x)\right)\left(\frac{d^2}{dx^2}y(x)\right)$$

and has the following solution:

> dsolve(",y(x));

$$y(x) = \_C1,\; x = 3/4\,\frac{y(x)^{4/3}}{\_C1} - \_C2$$

where _C1 and _C2 are constants of integration. The constant solution must be ignored.

Incorporating the boundary conditions (D26.6), we get

> u := subs(_C2=0,"[2]);

$$u := x = 3/4\,\frac{y(x)^{4/3}}{\_C1}$$

> subs({x=x1,y(x)=y1},"):solve(",{_C1}):s := subs(",u);

$$s := x = \frac{y(x)^{4/3} x1}{y1^{4/3}}$$

As the above s-expression shows, the extremizer is

$$y(x) = y_1(x/x_1)^{3/4} \qquad (D26.8)$$

Now one finds the minimal force from (D26.5):

```
> y := y1*(x/x1)^(3/4) :yx := diff(y,x):
> Fmin := 4*Pi*ro*v^2*int(yx^3*y,x=0..x1);
```

$$Fmin := \frac{27}{16} \frac{\text{Pi ro } v^2 y1^4}{x1^2}$$

where `ro` = ρ.

D27. Approximating boundary conditions

(MAPLE file vfem32)

The methods considered thus far assume the trial functions to satisfy the boundary conditions, while the residual of the governing equation or the functional is subject to minimization in one sense or another. An alternative way is to specify the trial functions so as to satisfy the equation and then minimize the residual in satisfying the boundary conditions in one sense or another (see Section D14). Consider the equation

$$T_{,xx} + T_{,yy} = 1 - x^2 \qquad (D27.1)$$

subjected to the boundary conditions

$$T(x=1,y) = T(x=-1,y) = T(x,y=1) = T(x,y=-1) = 0 \qquad (D27.2)$$

This equation may describe, among others, heat conduction in isotropic materials with internal sources, in which case T is the temperature.

Specify first a particular solution to (D27.1):

$$T_p = x^2/2 - x^4/12 \qquad (D27.3)$$

and then turn to the homogeneous (Laplace) equation for the function $T_h(x,y)$:

$$T^h_{,xx} + T^h_{,yy} = 0 \qquad (D27.4)$$

Obviously, $T(x,y) = T_p(x,y) + T_h(x,y)$.

The polynomials given by real or imaginary parts of the expression $(x + iy)^n$ provide a solution to (D27.4). Taking into account the symmetry of the problem, set a trial function as

$$T = x^2/2 - x^4/12 + c_0 + c_2 v_2(x,y) + c_4 v_4(x,y) + \cdots \qquad (D27.5)$$

where

$$v_2(x,y) = \text{Re}(x+iy)^2, \qquad v_4(x,y) = \text{Re}(x+iy)^4 \qquad (D27.6)$$

Since (D27.5) satisfies the governing equation (D27.1), the possible error is entirely due to the fact that it does not comply with the boundary conditions (D27.2). The coefficients c_0, c_2 and c_4 should be specified so as to minimize the residual in (D27.2). As earlier, a variety of residuals may be considered.

72 DIRECT METHODS

Let $T(c)$ and $v_i(c)$ be the values taken on by $T(x,y)$ and $v_i(x,y)$, respectively, at the boundary of the domain, C. One of the possible residuals to minimize is expressed by a set of equations of the Bubnov–Galerkin type

$$\int_C T(c)\,dc = 0$$

$$\int_C T(c)v_2\,dc = 0 \qquad (D27.7)$$

$$\int_C T(c)v_4\,dc = 0$$

Noting from (D27.5) and (D27.6) that the integrand in (D27.7) is an even function, these expressions simplify to

$$\int_0^1 T(x=1,y)\,dy + \int_0^1 T(x,y=1)\,dx = 0$$

$$\int_0^1 T(x=1,y)v_2(x=1,y)\,dy + \int_0^1 T(x,y=1)v_2(x,y=1)\,dx = 0 \qquad (D27.8)$$

$$\int_0^1 T(x=1,y)v_4(x=1,y)\,dy + \int_0^1 T(x,y=1)v_4(x,y=1)\,dx = 0$$

This provides three equations for specifying the three coefficients c_0, c_2 and c_4.

Going over to symbolic computations, first state v_2 and v_4 in agreement with (D27.6) and then (D27.5):

```
> w2 := Re((x+I*y)^2):w4 := Re((x+I*y)^4):
> v2 := evalc(w2);
                 2    2
         v2 := x  - y

> v4 := evalc(w4);
                 4      2  2    4
         v4 := x  - 6 x  y  + y

> T := x^2/2-x^4/12+c0+c2*v2+c4*v4;
               2         4              2    2         4      2  2    4
     T := 1/2 x  - 1/12 x  + c0 + c2 (x  - y ) + c4 (x  - 6 x  y  + y )
```

Verify whether the above function satisfies the equation (D27.1):

```
> eq := diff(T,x,x)+diff(T,y,y)-(1-x^2);
```

APPROXIMATING BOUNDARY CONDITIONS 73

$$eq := c4(12x^2 - 12y^2) + c4(-12x^2 + 12y^2)$$

> simplify(eq);

0

Now formulate (D27.8):

> R0 := int(subs(y=1,T),x=0..1)+int(subs(x=1,T),y=0..1);

$$R0 := -8/5\, c4 + \frac{17}{30} + 2\, c0$$

> R2 := int(subs(y=1,T*v2),x=0..1)+int(subs(x=1,T*v2), y=0..1);

$$R2 := \frac{16}{15}\, c2 + \frac{68}{315}$$

> R4 := int(subs(y=1,T*v4),x=0..1)+int(subs(x=1,T*v4), y=0..1);

$$R4 := -8/5\, c0 + \frac{1888}{315}\, c4 - \frac{614}{945}$$

Resolving this system of equations, specify the coefficients and then the function $T(x,y)$:

> solve({R0,R2,R4},{c0,c2,c4});

$$\{c0 = -1/4,\ c4 = 1/24,\ c2 = -\frac{17}{84}\}$$

> subs(",T);

$$\frac{28}{84}x^2 - 1/24\, x^4 - 1/4 + \frac{17}{84}y^2 - 1/4\, x^2 y^2 + 1/24\, y^4$$

One may improve the accuracy by considering the next approximation, given by

> v6 := evalc(Re((x+I*y)^6));

$$v6 := x^6 - 15\, x^4 y^2 + 15\, x^2 y^4 - y^6$$

> T := T+c6*v6;

$$T := 1/2\, x^2 - 1/12\, x^4 + c0 + c2(x^2 - y^2) + c4(x^4 - 6x^2 y^2 + y^4)$$
$$+ c6(x^6 - 15\, x^4 y^2 + 15\, x^2 y^4 - y^6)$$

and repeating the above procedure.

74 DIRECT METHODS

Another possible way of specifying the coefficients of the trial functions is the so-called *boundary collocation* method, according to which the residual $T(c)$ is set to zero at some boundary points. For example, for the third-order approximation one may set, instead of (D27.7),

$$T(x = 1, y = 0) = 0$$
$$T(x = 1, y = 1) = 0 \qquad \text{(D27.9)}$$
$$T(x = 0, y = 1) = 0$$

These ideas find further applications in the boundary element method, the treatment of which lies beyond the present scope.

D28. Use of logical expressions

Constructing trial functions is a crucial stage of applications of direct methods and various recommendations concerning this aspect have been made earlier. A resort to logical expressions and the technique of the so-called *R*-functions may also be useful.

Fig. D8 shows a geometrical interpretation of the three logical operations, namely, product of two regions (∩), sum of two regions (∪) and complement of a region (*). (Each region is considered as a set.) Fig. D8(a) illustrates the product of two sets A and B, Fig. D8(b) the sum of A and B and Fig. D8(c) the complement of A.

These and other logical operations may provide a basis for constructing trial functions which satisfy the prescribed boundary conditions.

As an example, consider a plate occupying the domain S with the boundary δ. Under proper boundary conditions, the problem may be reduced to finding a function v which possesses the following properties:

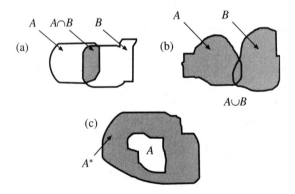

D8 Illustrating logical operations

$$v(\delta) = 0 \qquad \text{(a)}$$
$$v(S) > 0 \qquad \text{(b)} \qquad (D28.1)$$
$$v_{,n} = 1 \qquad \text{(c)}$$

where n is a unit normal to δ. If the S-domain has a complicated shape, selecting a trial function for v becomes a major problem. According to the above technique, this trial function may be constructed with the help of the v-functions for elementary 'brick' domains. These, in turn, follow from simple considerations. The last expression in (D28.1) is a normalization condition. If a v-function violates (D28.1c), it may be replaced by the associated normalized v_n-function given by

$$v = v/[v^2 + |\text{grad } v|^2]^{1/2} \qquad (D28.2)$$

Fig. D9 shows four basic domains, namely, circle, vertical strip, horizontal strip and ellipse, for which the v-functions are easily identified as

$$\begin{aligned} v &= a^2 - r^2 & \text{(a)} \\ v &= -ab + (a+b)x - x^2 & \text{(b)} \\ v &= -cd + (c+d)y - y^2 & \text{(c)} \\ v &= 1 - x^2/a^2 - y^2/b^2 & \text{(d)} \end{aligned} \qquad (D28.3)$$

Say, for an ellipse the normalized function v_n follows from (D28.2) as

$$v_n^{\text{ellipse}} = v/[v^2 + 4(x^2 + y^2/c^4)]^{1/2} \qquad (D28.4)$$

where v follows from (D28.3d) as

$$v = 1 - x^2 - y^2/c^2 \qquad (D28.5)$$

and $c = b/a$.

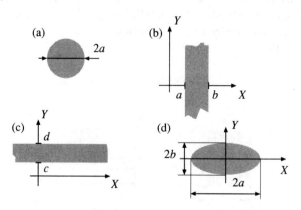

D9 Basic regions

76 DIRECT METHODS

Assume that a complicated domain S may be constructed by the above logical operations from simple basic domains. Then the 'algebraic' operations induced by these logical operations are

$$X \cap Y = X + Y - (X^2 + Y^2)^{1/2} \quad \text{(a) product}$$
$$X \cup Y = X + Y + (X^2 + Y^2)^{1/2} \quad \text{(b) sum} \quad \text{(D28.6)}$$
$$X^* = -X \quad \text{(c) complement}$$

where X and Y are the v-functions for the elementary 'brick' domains.

For example, for the square (see Fig. D10a) the logical expression is the product of the vertical and horisontal strips given by (D28.3b) and (D28.3c), respectively. On the other hand, the truncated ellipse (Fig. D10b) is the product of the ellipse and vertical strip.

Applying (D28.6a), we get for the square

$$v = v_{vs} + v_{hv} - (v_{vs}^2 + v_{hv}^2)^{1/2} \quad \text{(D28.7)}$$

where v_{vs} and v_{hv} are given by (D28.3b) and (D28.3c), respectively.

Assume that a clamped plate occupies the domain S. Then the deflection function $w(x,y)$ may suggest a representation

$$w(x,y) = v(x,y)f(x,y) \quad \text{(D28.8)}$$

where $v(x,y)$ is the above expression and $f(x,y)$ a Taylor's series. This technique applies also to other types of the boundary conditions.

D29. Concluding remarks

The above considerations show the potential of the direct methods in solving the problems of engineering analysis. Resorting to computer algebra further enhances the efficiency of these methods. In particular, it enables higher order approximations to be computed conveniently. Some of these methods may be reformulated so as to broaden their applicability. Note that the relevant rules must be adhered to in order to avoid erroneous results.

The next chapter presents finite element techniques closely related to the direct methods; Chapter W (Workshop) contains further examples of their application to problems of engineering analysis.

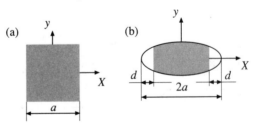

D10 Examples

F. Finite Element Method

The finite element approach brings in a new idea, namely, a spatial discretization of the domain under investigation. This allows us to further broaden the class of problems amenable to analysis, including those dealing directly with modern technology. On the other hand, this additional 'degree of freedom' may increase the danger of incorrect results. A sufficiently fine mesh and high order of approximating functions may ensure a reasonably small error.

Finite element techniques require the processing of extensive data and may efficiently be implemented only with the help of computers. That is why the method has been traditionally regarded as a purely numerical one. A resort to symbolic manipulation codes, like **MAPLE** or **MATHEMATICA**, may change the situation, allowing for an approximate but analytical solution.

Below we focus on the essence of this method, omitting mathematical technicalities, such as methods of integration or solution of systems of simultaneous equations, which may, at this stage, be left to the symbolic code.

In general, the presentation refers to results obtained via *vfem*-files with the extension .ms (MAPLE). To derive the relevant results with the help of MATHEMATICA the user must run the *vfem*-files with the extension .ma.

F1. Element stiffness matrix

(*MAPLE file vfem33*)

Fig. F1 shows a finite beam element subjected to the nodal forces q_1 and q_3 and the nodal moments q_2 and q_4. The corresponding nodal deflections are w_1 and w_3 and the corresponding nodal rotations are w_2 and w_4. These quantities enable one to define the work done on the element in the framework of elementary beam theory, which is the reason for their introduction. Note, the above q-forces are those acting *on* the nodes.

78 FINITE ELEMENT METHOD

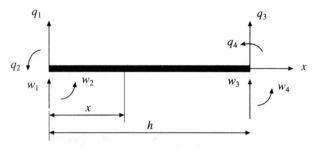

F1 Beam element

For convenience define the nodal displacement vector (matrix) $\{w\}$ and the nodal force vector (matrix) $\{q\}$, as follows:

$$\{w\} = [w_1, w_2, w_3, w_4]^T, \qquad \{q\} = [q_1, q_2, q_3, q_4]^T \tag{F1.1}$$

By definition, the element stiffness matrix relates $\{q\}$ and $\{w\}$ in a linear fashion:

$$\{q\} = [k]\{w\} \tag{F1.2}$$

In the case at hand $[k]$ is obviously a 4×4 matrix.

To construct the stiffness matrix begin with an approximation for the deflection w, which should contain four free coefficients, as the size of the $\{w\}$-matrix indicates. Indeed, the nodal quantities w_1, w_2, w_3, w_4 may be considered as the degrees of freedom of the system.

The simplest approximation is

$$w = a_0 + a_1 x + a_2 x^2 + a_3 x^3 \tag{F1.3}$$

which also specifies the rotation f as

$$f = dw/dx = w_{,xx} \tag{F1.4}$$

To derive the $[k]$, express the above a-coefficients in terms of $\{w\}$, observing that

$$w_1 = w(x=0), \quad w_3 = w(x=h), \quad w_2 = f(x=0), \quad w_4 = f(x=h) \tag{F1.5}$$

Solving (F1.5) provides the desirable dependencies, as the following MAPLE code shows (wx denotes $f = dw/dx$).

State first w and its derivative

```
> w:=sum(a[i]*x^i,i=0..3);
     w := a[0] + a[1]*x + a[2]*x**2 + a[3]*x**3
> wx:=diff(w,x);
     wx := a[1] + 2*a[2]*x + 3*a[3]*x**2
```

and then (F1.5)

ELEMENT STIFFNESS MATRIX

```
> d4:={w1=subs(x=0,w),w3=subs(x=h,w),f2=subs(x=0,wx),
f4=subs(x=h,wx)}:
```

Solving these for the *a*-coefficients, we get

```
> solution:=solve(d4,{a[0],a[1],a[2],a[3]});
```

solution := {a[0] = w1, a[1] = f2, a[3] = 1/h**3*(- 2*w3 + 2*w1
+ f2*h + h*f4), a[2] = - (h*f4 + 2*f2*h - 3*w3 + 3*w1)/h**2}

Substituting these back in *w* and *f*, we get these quantities in terms of the nodal displacement vector:

```
> subs(",w);
```

w1 + f2*x - (h*f4 + 2*f2*h - 3*w3 + 3*w1)/h**2*x**2 +
1/h**3*(- 2*w3 + 2*w1 + f2*h + h*f4)*x**3

```
> w:=";
```

w := w1 + f2*x - (h*f4 + 2*f2*h - 3*w3 + 3*w1)/h**2*x**2
+ 1/h**3*(- 2*w3 + 2*w1 + f2*h + h*f4)*x**3

```
> subs(solution,wx):
> wx:=";
```

wx := f2 - 2*(h*f4 + 2*f2*h - 3*w3 + 3*w1)/h**2*x
+ 3/h**3*(- 2*w3 + 2*w1 + f2*h + h*f4)*x**2

Note again that $wx = f = dw/dx$, $wxx = d^2w/dx^2$ and $wxxx = d^3w/dx^3$.
The elementary beam theory yields the following relations:

$$q_1 = EIw_{,xxx}(x=0), \qquad q_2 = -EIw_{,xx}(x=0),$$
$$q_3 = -EIw_{,xxx}(x=h), \qquad q_4 = EIw_{,xx}(x=h),$$
(F1.6)

which in fact define [*k*], as (F1.2) shows. It remains to program these expressions.

We first prepare the quantities appearing in the above expressions:

```
> wxx:=diff(wx,x);
```

wxx := - 2*(h*f4 + 2*f2*h - 3*w3 + 3*w1)/h**2
+ 6/h**3*(- 2*w3 + 2*w1 + f2*h + h*f4)*x

```
> wxxx:=diff(",x);
```

wxxx := 6/h**3*(- 2*w3 + 2*w1 + f2*h + h*f4)

```
> v[1]:=subs(x=0,wxxx):
> v[2]:=- subs(x=0,wxx):
> v[3]:=- subs(x=h,wxxx):
> v[4]:=subs(x=h,wxx):
```

80 FINITE ELEMENT METHOD

and then find the stiffness matrix $[k]$ making use of its definition, as given by (F1.2). In the MAPLE code the proper command is `genmatrix`. We temporarily neglect the multiplier EI.

```
> d10:=[v[1],v[2],v[3],v[4]];
```

    d10 := [6/h**3*(− 2*w3 + 2*w1 + f2*h + h*f4), 2*(h*f4 + 2*f2*h − 3*w3 + 3*w1)/h**2, − 6/h**3* (− 2*w3 + 2*w1 + f2*h + h*f4), − 2*(h*f4 + 2*f2*h − 3*w3 + 3*w1)/h**2 + 6/h**2*(− 2*w3 + 2*w1 + f2*h + h*f4)]

```
> with(linalg,genmatrix):
> kk:=genmatrix(d10,[w1,f2,w3,f4]);
```

    kk := MATRIX(((12/h**3, 6/h**2, −12/h**3, 6/h**2], [6/h**2, 4/h, − 6/h**2, 2/h], [−12/h**3, − 6/h**2, 12/h**3, − 6/h**2], [6/h**2, 2/h, − 6/h**2, 4/h]])

It remains to recover the bending stiffness EI to find $[k]$:

```
> k:=evalm(EI*kk);
```

    k := MATRIX(((12*EI/h**3, 6*EI/h**2, −12*EI/h**3, 6*EI/h**2], [6*EI/h**2, 4*EI/h, −6*EI/h**2, 2*EI/h], [−12*EI/h**3, −6*EI/h**2, 12*EI/h**3, −6*EI/h**2], [6*EI/h**2, 2*EI/h, −6*EI/h**2, 4*EI/h]])

For convenience, we present the stiffness matrix $[k]$ in the two-dimensional format (prettyprint) too:

$$\begin{bmatrix} 12\dfrac{EI}{h^3} & 6\dfrac{EI}{h^2} & -12\dfrac{EI}{h^3} & 6\dfrac{EI}{h^2} \\ 6\dfrac{EI}{h^2} & 4\dfrac{EI}{h} & -6\dfrac{EI}{h^2} & 2\dfrac{EI}{h} \\ -12\dfrac{EI}{h^3} & -6\dfrac{EI}{h^2} & 12\dfrac{EI}{h^3} & -6\dfrac{EI}{h^2} \\ 6\dfrac{EI}{h^2} & 2\dfrac{EI}{h} & -6\dfrac{EI}{h^2} & 4\dfrac{EI}{h} \end{bmatrix}$$

Note that this matrix is symmetric.

The deflection $w(x)$ as given by (F1.3) may be stated in terms of the nodal displacement matrix $\{w\}$ with the help of the so-called *interpolation* (*shape*)

ELEMENT STIFFNESS MATRIX

functions. To this end, set

$$w(x) = \{d(x)\}^T \{w\} \tag{F1.7}$$

where

$$\{d(x)\}^T = [d_1(x)\, d_2(x)\, d_3(x)\, d_4(x)] \tag{F1.8}$$

is the as yet unknown matrix of interpolation functions.

As (F1.7) shows, to find the explicit form of these functions one may use the command coeff, as follows:

```
> d[1]:=coeff(collect(w,w1),w1,1);
    d[1]:= - 3/h**2*x**2 + 1 + 2/h**3*x**3
> d[2]:=coeff(collect(w,f2),f2,1);
    d[2]:= x - 2/h*x**2 + 1/h**2*x**3
> d[3]:=coeff(collect(w,w3),w3,1);
    d[3]:= 3/h**2*x**2 - 2/h**3*x**3
> d[4]:=coeff(collect(w,f4),f4,1);
    d[4]:= - 1/h*x**2 + 1/h**2*x**3
```
(F1.9)

The interpolation functions are of major interest in the finite element method and will be of frequent use in subsequent considerations. For the purpose of explicitness, we present also their prettyprint format:

```
> d[1]:=coeff(collect(w,w1),w1,1);
```
$$d[1] := -3\frac{x^2}{h^2} + 1 + 2\frac{x^3}{h^3}$$

```
> d[2]:=coeff(collect(w,f2),f2,1);
```
$$d[2] := x - 2\frac{x^2}{h} + \frac{x^3}{h^2}$$

```
> d[3]:=coeff(collect(w,w3),w3,1);
```
$$d[3] := 3\frac{x^2}{h^2} - 2\frac{x^3}{h^3}$$

```
> d[4]:=coeff(collect(w,f4),f4,1);
```
$$d[4] := \frac{-x^2}{h} + \frac{x^3}{h^2}$$

Note the following basic property of the interpolation (shape) functions. If x_i is one of the nodes and $d_j(x)$ one of the shape functions, then

$$d_j(x = x_i) = \delta_{ij} \tag{F1.10}$$

as can be seen from (F1.7).

As (F1.7) and (F1.9) show, the state of the finite element is completely specified provided $\{w\}$ is given. This is why the size of $\{w\}$ defines the number of degrees of freedom of the element. In the case at hand, there are four degrees of freedom.

One may conclude that (1) the nodal quantities $\{q\}$ and $\{w\}$ may indeed be related by (F1.2) with $[k]$ being the stiffness matrix, (2) the field variable w may be expressed in terms of the nodal vector $\{w\}$ by (F1.7) where $\{d(x)\}$ is the vector of shape functions. The representations similar to (F1.2) and (F1.7) are typical of finite element analysis.

F2. Energy analysis

(MAPLE file vfem33)

Previous considerations indicate that the coefficients of the approximating polynomial or the nodal displacement vector $\{w\}$ may serve as the generalized degrees of freedom of the beam element. The kinetic and potential energies of this element must therefore be expressible in terms of $\{w\}$.

Considering a dynamic bending and setting $\{w\} = \{w(t)\}$, we state the kinetic energy of the element:

$$T = 1/2 \int_0^h \rho(x) w_{,t}^2(x, t)\, dx \tag{F2.1}$$

where $\rho(x)$ is the mass density per unit length. Equation (F1.7) yields

$$\{w_{,t}\}^T \{w_{,t}\} = w_{,t}^2 = \{w_{,t}(t)\}^T \{d(x)\}\{d(x)\}^T \{w_{,t}(t)\} \tag{F2.2}$$

This provides for the kinetic energy

$$T = 1/2 \{w_{,t}(t)\}^T [m] \{w_{,t}(t)\} \tag{F2.3}$$

with the *consistent* mass matrix $[m]$ given by

$$[m] = \int_0^h \{d(x)\} \rho(x) \{d(x)\}^T dx \tag{F2.4}$$

ENERGY ANALYSIS   83

In a completely similar way, the potential energy U is

$$U = 1/2 \int_0^h EI(x) w_{,xx}^2 \, dx \qquad (F2.5)$$

Differentiating (F1.7) twice with respect to x and using the result in (F2.5), we get

$$U = 1/2\{w(t)\}^T [k]\{w(t)\} \qquad (F2.6)$$

with the stiffness matrix $[k]$ given by

$$[k] = \int_0^h \{d_{,xx}(x)\} EI(x)\{d_{,xx}(x)\}^T \, dx \qquad (F2.7)$$

Note that (F2.3) and (F2.6) are remarkably similar to the kinetic energy of a particle and the potential energy of the spring, respectively.

Going over to the external load applied to the element, note that in general it may be represented with the help of its intensity function, say, $s(x, t)$. A question arises about equivalent representations of $s(x, t)$ in terms of the nodal forces.

To this end, we get from (F1.7) the virtual deflection $\delta w(x, t)$

$$\delta w(x, t) = \{d(x)\}^T \{\delta w(t)\} \qquad (F2.8)$$

which provides the virtual work done on this deflection by the external forces:

$$\int_0^h s(x, t) \delta w \, dx = \{p(t)\}^T \delta\{w(t)\} \qquad (F2.9)$$

where the equivalent nodal forces matrix $\{p(t)\}$ is given by

$$\{p(t)\}^T = \int_0^h s(x, t)\{d(x)\}^T \, dx \qquad (F2.10)$$

Using the delta function, one may apply (F2.10) to describe the concentrated forces too. Eqs. (F2.4) and (F2.7) show that the elemental mass- and stiffness matrices are symmetric.

As a sequel to the program of the previous section, we compute below $[m]$, denoting ρ by r and using (F2.4):

```
> m:=array(1..4,1..4):
> for i to 4 do
> for j to 4 do
> if i<=j then
> m[i,j]:=r*int((d[i]*d[j]),x=0..h)
> else m[i,j]:=m[j,i]
```

84 FINITE ELEMENT METHOD

```
> fi:
> od:
> od:
> print(m);
```

$$\begin{bmatrix} \dfrac{13}{35}rh & \dfrac{11}{210}rh^2 & 9/70\,rh & -\dfrac{13}{420}rh^2 \\ \dfrac{11}{210}rh^2 & 1/105\,rh^3 & \dfrac{13}{420}rh^2 & -1/140\,rh^3 \\ 9/70\,rh & \dfrac{13}{420}rh^2 & \dfrac{13}{35}rh^2 & -\dfrac{11}{210}rh^2 \\ -\dfrac{13}{420}rh^2 & -1/140\,rh^3 & -\dfrac{11}{210}rh^2 & 1/105\,rh^3 \end{bmatrix}$$

We note in passing that besides the above consistent mass matrix, one may introduce the diagonal or *lumped* mass matrices, which are obtained by placing particle masses m_j at nodes j of an element in a way that ensures the correct total mass. Though these matrices may perform well in particular problems, they will not be considered in this text.

A completely similar procedure provides the stiffness matrix $[k]$ via (F2.7)

```
> for i to 4 do
> for j to 4 do
> if i<=j then
> k[i,j]:=EI*int(diff(d[i],x$2)*diff(d[j],x$2),x=0..h)
> else k[i,j]:=k[j,i]
> fi:
> od:
> od:
> print(k);
```

$$\begin{bmatrix} 12\dfrac{EI}{h^3} & 6\dfrac{EI}{h^2} & -12\dfrac{EI}{h^3} & 6\dfrac{EI}{h^2} \\ 6\dfrac{EI}{h^2} & 4\dfrac{EI}{h} & -6\dfrac{EI}{h^2} & 2\dfrac{EI}{h} \\ -12\dfrac{EI}{h^3} & -6\dfrac{EI}{h^2} & 12\dfrac{EI}{h^3} & -6\dfrac{EI}{h^2} \\ 6\dfrac{EI}{h^2} & 2\dfrac{EI}{h} & -6\dfrac{EI}{h^2} & 4\dfrac{EI}{h} \end{bmatrix}$$

Now consider an example involving a distributed external force $s(x, t)$, say, the case $s(x,t) = \text{const} = e_0$. Then (F2.10) straightforwardly provides

$$\{p\} = e_0 \int_0^h \{d(x)\}\, dx \qquad (F2.11)$$

This takes the following explicit form in the MAPLE code ($e0 = e_0$),

```
>p:=array (1..4,1..1,[]):
> for i to 4 do
>p[i,1]:=e0*int(d[i],x=0..h)
> od:
> print (p);
```

$$\begin{bmatrix} 1/2\,e0\,h \\ 1/12\,e0\,h^2 \\ 1/2\,e0\,h \\ -1/12\,e0\,h^2 \end{bmatrix}$$

F3. Truss element

(MAPLE file vfem33)

Fig. F2 shows a truss consisting of three pin-connected bars, each of which may be thought of as a finite element. The element suffers a purely axial stress due to the axial displacement u, since no bending takes place. Nevertheless, it may undergo a transverse displacement v, which is a rigid body motion. This displacement contributes nothing to the strain energy, but does influence the kinetic energy. Each of the elements may be referred to either the local

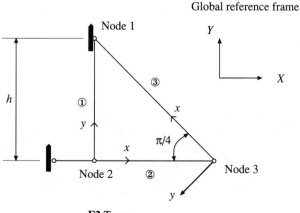

F2 Truss

86 FINITE ELEMENT METHOD

coordinate system with the x-coordinate directed along the element axis or to a global coordinate system.

The bar element has four degrees of freedom, as each of the nodes has two. Therefore, approximations for the displacement u and displacement v, which contain four coefficients, would be appropriate:

$$u(x) = a + a_1 x, \qquad v(x) = b_0 + b_1 x \qquad (F3.1)$$

where x is the axial coordinate. Now introduce the nodal displacement 4×1 matrix $\{w\}$:

$$\{w\} = [w_1 w_2 w_3 w_4]^T \qquad (F3.2)$$

with $w_1 = u_1$, $w_2 = v_1$, $w_3 = u_2$, $w_4 = v_2$, as shown in Fig. F3. Following the procedure of the previous section, express the coefficients appearing in (F3.1) in terms of the nodal displacement vector $\{w\}$.

To this end, substituting $x = 0$ and then $x = h$, we get from (F3.1)

$$a_1 = (u_2 - u_1)/h, \qquad a_0 = u_1, \qquad b_1 = (v_2 - v_1)/h, \qquad b_0 = v_1 \qquad (F3.3)$$

Equation (F3.1) acquires the form

$$u(x) = (1 - x/h)u_1 + (x/h)u_2, \qquad v(x) = (1 - x/h)v_1 + (x/h)v_2 \qquad (F3.4)$$

Introducing the notation for the interpolation (shape) functions,

$$d_1(x) = 1 - x/h, \qquad d_2(x) = x/h \qquad (F3.5)$$

we get the field variables:

$$u(x) = d_1(x)u_1 + d_2(x)u_2, \qquad v(x) = d_1(x)v_1 + d_2(x)v_2 \qquad (F3.6)$$

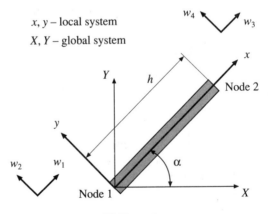

F3 Truss element

or, in the familiar matrix form,

$$u(x) = \{w\}^T\{D_1(x)\}, \qquad v(x) = \{w\}^T\{D_2(x)\} \qquad (F3.7)$$

Here use has been made of (F3.2) and the notation

$$\{D_1(x)\} = [d_1(x), 0, d_2(x), 0]^T, \qquad \{D_2(x)\} = [0, d_1(x), 0, d_2(x)]^T \qquad (F3.8)$$

Now the kinetic energy takes the form

$$T = 1/2 \int_0^h \rho(x)[u_{,t}^2(x,t) + v_{,t}^2(x,t)]\,dx$$

$$= 1/2 \int_0^h \rho(x)\{w_{,t}\}^T[\{D_1(x)\}\{D_1(x)\}^T + \{D_2(x)\}\{D_2(x)\}^T]\{w_{,t}\}\,dx \qquad (F3.9)$$

This provides the following elemental mass matrix $[m]$:

$$[m] = \int_0^h \rho(x)[\{D_1(x)\}\{D_1(x)\}^T + \{D_2(x)\}\{D_2(x)\}^T]\,dx \qquad (F3.10)$$

and the kinetic energy of the truss element acquires a form identical to that of the beam element, as given by (F2.3).

As to the potential energy U, only the axial displacement u comes into play. Considering, say, the element 2, get

$$U = 1/2 \int_0^h EA(x)u_{,x}^2(x,t)\,dx \qquad (F3.11)$$

with $EA(x)$ being the variable stiffness. In view of (F3.7), get

$$U = 1/2 \int_0^h \{w\}^T\{D_{1,x}(x)\}EA(x)\{D_{1,x}(x)\}^T\{w\}\,dx = 1/2\{w\}^T[k]\{w\}\,dx \qquad (F3.12)$$

with the elemental stiffness matrix $[k]$ given by

$$[k] = 1/2 \int_0^h \{D_{1,x}(x)\}EA(x)\{D_{1,x}(x)\}^T\,dx \qquad (F3.13)$$

The right-hand side of the expression (F3.12) is identical to the expression for the strain energy of the beam element, obtained earlier.

88　FINITE ELEMENT METHOD

Denoting D1 = $\{D_1(x)\}$, D2 = $\{D_2(x)\}$ and r = ρ, we compute below $[m]$ and then $[k]$. First, state the shape functions:

```
> with(linalg,multiply,transpose,add):
> D1:=array([[1-x/h],[0],[x/h],[0]]);
```

$$D1 := \begin{bmatrix} 1 - x/h \\ 0 \\ x/h \\ 0 \end{bmatrix}$$

```
> D2:=array([[0],[1-x/h],[0],[x/h]]);
```

$$D2 := \begin{bmatrix} 0 \\ 1 - x/h \\ 0 \\ x/h \end{bmatrix}$$

The bracketed term in (F3.10) is denoted below as d4:

```
> multiply(D1,transpose(D1)):
> multiply(D2,transpose(D2)):
> d4:=add(","");
```

d4 := MATRIX(((($-h+x$)**2/h**2, 0, $-(-h+x)$*x/h**2, 0],
[0, ($-h+x$)**2/h**2, 0, $-(-h+x)$*x/h**2], [$-(-h+x)$*x/h**2,
0, 1/h**2*x**2, 0], [0, $-(-h+x)$*x/h**2, 0, 1/h**2*x**2]])

This enables us, using (F3.10), to compute mbar = $[m]$ for the case of a uniform element:

```
> mbar:=array(1..4,1..4,[]):
> for i to 4 do
> for j to 4 do
> if i<=j then
> mbar[i,j]:=r*int(d4[i,j],x=0..h)
> else mbar[i,j]:=mbar[j,i]
> fi:
> od
> od:
> print(mbar);
```

```
MATRIX(((1/3*r*h, 0, 1/6*r*h, 0], [0, 1/3*r*h, 0, 1/6*r*h],
[1/6*r*h, 0, 1/3*r*h, 0], [0, 1/6*r*h, 0, 1/3*r*h]])
```

For convenience, this matrix is also given in the prettyprint mode

```
> print(mbar);
```

$$\begin{bmatrix} 1/3\,rh & 0 & 1/6\,rh & 0 \\ 0 & 1/3\,rh & 0 & 1/6\,rh \\ 1/6\,rh & 0 & 1/3\,rh & 0 \\ 0 & 1/6\,rh & 0 & 1/3\,rh \end{bmatrix}$$

Going over to the stiffness matrix kbar = $[k]$, we denote D1x = $\{D_{1,x}(x)\}$ and compute the integrand of (F3.13):

```
> D1x:=array(1..4,1..1,[]):
> for i to 4 do
> for j to 1 do
> D1x[i,j]:=diff(D1[i,j],x)
> od:
> od:
> print(D1x);
```

MATRIX(((− 1/h], [0], [1/h], [0]])

```
> d6:=multiply(D1x,transpose(D1x)):
```

The stiffness matrix kbar = $[k]$ is

```
> kbar:=array(1..4,1..4):
> for i to 4 do
> for j to 4 do
> if i<=j then
> kbar[i,j]:=EA*int(d6[i,j],x=0..h)
> else kbar[i,j]:=kbar[j,i]
> fi:
> od:
> od:
> print(kbar);
```

MATRIX(((EA/h, 0, − EA/h, 0], [0, 0, 0, 0], [− EA/h, 0, EA/h, 0], [0, 0, 0, 0]])

The prettyprint format of $[k]$ is

```
> print(kbar);
```

$$\begin{bmatrix} \frac{EA}{h} & 0 & -\frac{EA}{h} & 0 \\ 0 & 0 & 0 & 0 \\ -\frac{EA}{h} & 0 & \frac{EA}{h} & 0 \\ 0 & 0 & 0 & 0 \end{bmatrix}$$

90 FINITE ELEMENT METHOD

The results obtained apply to each of the bars in its local cooordinate system.

F4. Physical meaning of matrices

(MAPLE file vfem33)

The stiffness matrix $[k]$ relates the nodal force matrix $\{q\}$ to the nodal displacement matrix $\{w\}$, as given by (F1.2). More precisely, the k_{ij}-entry of $[k]$ is equal to the nodal force in the i-direction under the unit displacement in the j-direction. To illustrate this, assume that $\{w\}^T = [1, 0, 0, 0]$ (which means that the deflection at the node 1 is unity and all other components of the displacement vanish) and find the associated nodal forces $\{q\}$ for the beam element.

Continuing the program of the previous sections and using the notation w = $\{w\}$ and q = $\{q\}$, we get once again the stiffness matrix of the beam element

```
>print(k);
```

   MATRIX(((12*EI/h**3, 6*EI/h**2, − 12*EI/h**3, 6*EI/h**2], [6*EI/
   h**2, 4*EI/h, − 6*EI/h**2, 2*EI/h], [− 12*EI/h**3, − 6*EI/h**2,
   12*EI/h**3, − 6*EI/h**2], [6*EI/h**2, 2*EI/h, − 6*EI/h**2, 4*EI/h]])

specify $\{w\}$

```
>w:=array([[1],[0],[0],[0]]);
```

   w := MATRIX(((1], [0], [0], [0]])

and find $\{q\}$

```
>q:=multiply(k,w);
```

   q := MATRIX(((12*EI/h**3], [6*EI/h**2], [−12*EI/h**3], [6*EI/h**2]])

It is observed that $\{q\}$ is merely the first column of $[k]$ with the components describing, respectively, the transverse forces and moments at the nodes under the unit displacement in the transverse direction.

So far we have invoked no boundary conditions. This means that the results obtained apply to a structure, which is free to move in space, the so-called 'floating' structure. This manifests in the determinant of $[k]$. Indeed,

```
>with(linalg,det):
>det(k);
```

   0

PHYSICAL MEANING OF MATRICES 91

Thus, [k] is a singular matrix. In general, the vanishing determinant is typical of the structures subjected to a rigid body motion. Such a motion gives rise to zero nodal forces. For example, we may specify a pure translation of the beam element to the value z and find that the associated nodal forces $\{q\}$ vanish:

```
> w:=array([[z],[0],[z],[0]]);
```
   $w := \text{MATRIX}(((z], [0], [z], [0]])$

```
> q:=multiply(k,w);
```
   $q := \text{MATRIX}(((0], [0], [0], [0]])$

The mass matrix [m] describes the inertial properties of the element, more precisely the 'distribution' of the inertia over the degrees of freedom of the element. Without loss of generality, assume the nodal displacement to vary harmonically in time with the unit amplitude and compute the associated inertial forces. Below w1, wt2 and i denote the frequency, acceleration and the inertial forces, respectively:

```
> w:=array([[1*sin(w1*t)],[0],[0],[0]]);
```
   $w := \text{MATRIX}(((\sin(w1^*t)], [0], [0], [0]])$

```
> wt2:=array(1..4,1..1,[]):
> for i to 4 do
> wt2[i,1]:=diff(w[i,1],t$2)
> od:
> print(wt2);
```
   $\text{MATRIX}(((-\sin(w1^*t)^*w1^{**}2], [0], [0], [0]])$

```
> print(m);
```
   MATRIX(((13/35*r*h, 11/210*r*h**2, 9/70*r*h, −13/420*r*h**2],
   [11/210*r*h**2, 1/105*r*h**3, 13/420*r*h**2, −1/140*r*h**3],
   [9/70*r*h, 13/420*r*h**2, 13/35*r*h, −11/210*r*h**2],
   [−13/420*r*h**2, −1/140*r*h**3, −11/210*r*h**2, 1/105*r*h**3]])

```
> i:=multiply(m,wt2);
```
   $i := \text{MATRIX}(((-13/35^*r^*h^*\sin(w1^*t)^*w1^{**}2],$
   [−11/210*r*h**2*sin(w1*t)*w1**2], [−9/70*r*h*sin(w1*t)*w1**2],
   [13/420*r*h**2*sin(w1*t)*w1**2]])

Thus, the $i_{\alpha\beta}$-entry of the mass matrix i is equal to the nodal inertia force in the α-direction under the unit harmonic displacement in the β-direction. We may

92 FINITE ELEMENT METHOD

conclude that the matrices $[m]$ and $[k]$ enable one to conveniently model such fundamental properties of the 'piece' of continuum as inertia and elasticity by those of a system with a limited number of degrees of freedom. This is a remarkable feature of the finite element method.

F5. Eigenvalues of stiffness matrix

(*MAPLE file vfem33*)

Invariants of a matrix, its eigenvalues in particular, usually bear essential physical information. Denote the eigenvalue as λ and state the eigenvalue problem as

$$([k] - \lambda[Id])\{w\} = 0 \qquad (F5.1)$$

where $[Id]$ is the unit matrix and $[k]$ the stiffness matrix. For $\lambda = 0$, we trivially get

$$[k]\{w\} = \{q\} = 0 \qquad (F5.2)$$

which indicates that this case corresponds to vanishing nodal forces. This, in turn, may suggest a rigid body motion of the element, as considerations of the previous section show. The associated eigenvector describes the type of motion.

When $\lambda \neq 0$, the eigenvector describes a mode of the strain state of the element. Illustrating these comments, we compute below the eigenvalues of the stiffness matrix of the beam element.

Note that the entries of $[k]$ may have different dimensions. In the case of the beam, it is convenient to modify the basic relation (F1.2) to

$$\{q_h\} = [k_h]\{w_h\} \qquad (F5.3)$$

where

$$\{q_h\} = [q_1 \ q_2/h \ q_3 \ q_4/h]^T, \quad \{w_h\} = [w_1 \ w_2 h \ w_3 \ w_4 h]^T \qquad (F5.4)$$

and the stiffness matrix $[k_h] = $ kh is given below:

```
> kh:=evalm(EI/h^3*array([[12,6,-12,6],[6,4,-6,2],
  [-12,-6,12,-6],
> [6,2,-6,4]]));
```

EIGENVALUES OF STIFFNESS MATRIX

$$kh := \begin{bmatrix} 12\dfrac{EI}{h^3} & 6\dfrac{EI}{h^3} & -12\dfrac{EI}{h^3} & 6\dfrac{EI}{h^3} \\ 6\dfrac{EI}{h^3} & 4\dfrac{EI}{h^3} & -6\dfrac{EI}{h^3} & 2\dfrac{EI}{h^3} \\ -12\dfrac{EI}{h^3} & -6\dfrac{EI}{h^3} & 12\dfrac{EI}{h^3} & -6\dfrac{EI}{h^3} \\ 6\dfrac{EI}{h^3} & 2\dfrac{EI}{h^3} & -6\dfrac{EI}{h^3} & 4\dfrac{EI}{h^3} \end{bmatrix}$$

with all the entries having the same dimension.

Going over to the eigenvalues and continuing the previous program, we get

```
> with(linalg,eigenvals):
> eig:=eigenvals(kh);
```

$$eig := 0, 0, 30\dfrac{EI}{h^3}, 2\dfrac{EI}{h^3}$$

The first two eigenvalues vanish, which may correspond to a rigid body motion and zero nodal forces, as noted earlier.

Verifying this point, denote temporarily the nodal forces as b and compute them for an arbitrary nodal displacement $wh = \{w_h\}$

```
> wh:=array([[w1],[w2h],[w3],[w4h]]);
```

$$wh := \begin{bmatrix} w1 \\ w2h \\ w3 \\ w4h \end{bmatrix}$$

```
> b:=multiply(kh,wh);
```

$$b := \begin{bmatrix} 6\,\dfrac{EI(2w1 + w2h - 2w3 + w4h)}{h^3} \\ 2\,\dfrac{EI(3w1 + 2w2h - 3w3 + w4h)}{h^3} \\ -6\,\dfrac{EI(2w1 + w2h - 2w3 + w4h)}{h^3} \\ 2\,\dfrac{EI(3w1 + w2h - 3w3 + 2w4h)}{h^3} \end{bmatrix}$$

94 FINITE ELEMENT METHOD

For zero eigenvalue, we get from (F5.2) that
$$b = \{b\} = 0 \tag{F5.5}$$

Therefore, solving (F5.5) for $\{w_h\}$, get the associated displacement weigen:

```
> weigen:=solve({b[1,1],b[2,1],b[3,1],b[4,1]},
  {w1,w2h,w3,w4h});
```

weigen := $\{w1 = w3 - w4h, w2h = w4h, w3 = w3, w4h = w4h\}$

The obtained values of $\{w\}$ indeed describe the rigid body motion consisting of rotation to the angle $w_2h = w_4h$ and translation of the two nodes to the values w_3 and $(w_3 - w_4h)$, respectively.

Turning to non-zero eigenvalues and denoting the unit matrix as $\text{Id} = [Id]$, compute first the bracketed term of (F5.1) for the third eigenvalue

```
> Id:=array(identity,1..4,1..4);
```

Id := array(identity,1 .. 4,1 .. 4,[])

```
> evalm(kh-eig[3]*Id);
```

$$\begin{bmatrix} -18\frac{EI}{h^3} & 6\frac{EI}{h^3} & -12\frac{EI}{h^3} & 6\frac{EI}{h^3} \\ 6\frac{EI}{h^3} & -26\frac{EI}{h^3} & -6\frac{EI}{h^3} & 2\frac{EI}{h^3} \\ -12\frac{EI}{h^3} & -6\frac{EI}{h^3} & -18\frac{EI}{h^3} & -6\frac{EI}{h^3} \\ 6\frac{EI}{h^3} & 2\frac{EI}{h^3} & -6\frac{EI}{h^3} & 4\frac{EI}{h^3} \end{bmatrix}$$

and then multiply this by the nodal displacement vector $\{w_h\}$ = wh

```
> d:=multiply(",wh);
```

d := MATRIX(((-6*EI*(3*w1 - w2h + 2*w3 - w4h)/h**3],
[2*EI*(3*w1 - 13*w2h-3*w3 + w4h)/h**3],
[- 6*EI*(2*w1 + w2h + 3*w3 + w4h)/h**3],
[2*EI*(3*w1 + w2h - 3*w3 - 13*w4h)/h**3]])

Solving this for the components of $\{w\}$, get the associated eigenvector

```
> solve({d[1,1],d[2,1],d[3,1],d[4,1]},{w1,w2h,w3,w4h});
```

$\{w2h = w4h, w4h = w4h, w3 = -2*w4h, w1 = 2*w2h\}$

Thus, this strain mode consists of the equal but opposite displacement of the nodes and their arbitrary identical rotation. The associated nodal forces $\{q_h\}$ are

```
>weigen:=array([[2*w2h],[w2h],[-2*w2h],[w2h]]):
>qh:=multiply(kh,weigen);
```

$$qh := MATRIX(((60*EI/h**3*w2h], [30*EI/h**3*w2h],$$
$$[-60*EI/h**3*w2h], [30*EI/h**3*w2h]])$$

Repeating this scheme for the fourth eigenvalue, we obtain

```
>evalm(kh-eig[4]*Id):
>d:=multiply(",wh);
```

$$d := MATRIX(((2*EI*(5*w1 + 3*w2h - 6*w3 + 3*w4h)/h**3],$$
$$[2*EI*(3*w1 + w2h - 3*w3 + w4h)/h**3],$$
$$[-2*EI*(6*w1 + 3*w2h - 5*w3 + 3*w4h)/h**3],$$
$$[2*EI*(3*w1 + w2h - 3*w3 + w4h)/h**3]])$$

```
>solve({d[1,1],d[2,1],d[3,1],d[4,1]},{w1,w2h,w3,w4h});
```

$$\{w1 = 0, w4h = w4h, w3 = 0, w2h = -w4h\}$$

Thus, this eigenvector describes rotation of the nodal cross-sections to the equal opposite angles. The associated nodal force vector $\{q_h\}$ consists of two moments:

```
>weigen:=array([[0],[w2h],[0],[-w2h]]);
```

$$weigen := MATRIX(((0], [w2h], [0], [-w2h]])$$

```
>qh:=multiply(kh,weigen);
```

$$qh := MATRIX(((0], [2*EI/h**3*w2h], [0], [-2*EI/h**3*w2h]])$$

F6. Reference systems

(MAPLE file vfem34)

The finite element method suggests that a structure or domain of interest may be represented as an assemblage of finite elements, each of which is amenable to analysis in its own local reference frame. Since the orientation of elements may be different, one usually needs to perform a transformation of relevant matrices to a single global reference frame.

Referring, for example, to a bar element, we note that in space it has six degrees of freedom, namely w_1, w_2, ..., w_6, in the local coordinate system x, y, z and six degrees of freedom, namely W_1, W_2, ..., W_6, in the global coordinate system, X, Y, Z. Indeed, the bar element has two nodes, each having three degrees of freedom. The relevant transformation is given by

$$[x, y, z]^T = [n][X, Y, Z]^T \tag{F6.1}$$

with $[n] = [n_{ij}]$, $i, j = 1, 2, 3$ being the matrix of direction cosines. Say, n_{11} is the cosine of the angle between the X- and x-axes. The same transformation relates the displacement vector of the node

$$[w_1 \, w_2 \, w_3]^T = [n][W_1 \, W_2 \, W_3]^T, \quad [w_4 \, w_5 \, w_6] = [n][W_4 \, W_5 \, W_6]^T \tag{F6.2}$$

Introducing the 'total' vectors of the nodal displacement in the local and global coordinates, $\{w\}$ and $\{W\}$, respectively:

$$\{w\} = [w_1 \, w_2 \, w_3 \, w_4 \, w_5 \, w_6]^T, \quad \{W\} = [W_1 \, W_2 \, W_3 \, W_4 \, W_5 \, W_6]^T \tag{F6.3}$$

we get

$$\{w\} = [N]\{W\} \tag{F6.4}$$

Here the N-matrix is

$$[N] = \begin{bmatrix} [n] & 0 \\ 0 & [n] \end{bmatrix} \tag{F6.5}$$

The above matrices $[n]$ and $[N]$ are orthonormal, as they describe a rigid body rotation. Their explicit form will be given later. The vector of distributed force $\{p\}$ transforms in a similar way:

$$\{p\} = [N]\{P\} \tag{F6.6}$$

Now the expressions for the kinetic and strain energies can be written in the global coordinate system too. For the kinetic energy, we get from (F2.3)

$$T = 1/2\{w_{,t}\}^T[m]\{w_{,t}\} = 1/2\{W_{,t}\}^T[N]^T[m][N]\{W_{,t}\} = 1/2\{W_{,t}\}^T[M]\{W_{,t}\} \tag{F6.7}$$

and for the strain energy we get from (F2.6)

$$U = 1/2\{w\}^T[k]\{w\} = 1/2\{W\}^T[N]^T[k][N]\{W\} = 1/2\{W\}^T[K]\{W\} \tag{F6.8}$$

with the obvious notation for the elemental mass matrix in the global system

$$[M] = [N]^T[m][N] \tag{F6.9}$$

and for the elemental stiffness matrix in the global system

$$[K] = [N]^T[k][N] \tag{F6.10}$$

Now we specify the matrix $[n]$ appearing in the above for the planar case. Fig. F4 shows a local and global coordinate systems, for both of which the shortest rotation from the first axis (x or X) to the second (y or Y) occurs counterclockwise.

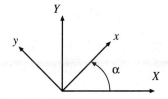

F4 Local and global reference systems

For such reference frame the transformation matrix $[n]$ is given by

$$[n] := \begin{bmatrix} \cos(\alpha) & \sin(\alpha) & 0 \\ -\sin(\alpha) & \cos(\alpha) & 0 \\ 0 & 0 & 1 \end{bmatrix} \quad \text{(F6.11)}$$

If X_1 and X_2 are the X-coordinates of the nodes 1 and 2, then $\cos\alpha = (X_2 - X_1)/h$, etc. This makes it possible to construct the matrix $[N]$ via (F6.5) and then $[M]$ and $[K]$ via (F6.9) and (F6.10), respectively.

As an illustration, consider the bar 3 of the truss shown in Fig. F2. The angle $\alpha = 3\pi/4$ and the corresponding matrix $[n]$ follows from (F6.11). Below we compute $[M]$ and $[K]$ for this bar, denoting these quantities as M3 and K3, respectively. To this end, first specify $[N]$, then $[m]$ and $[k]$, as they were found earlier. In doing so, adjust the length to be $2^{1/2}h$.

```
> with(linalg,multiply,transpose):
> d1:=array([[-1,1,0,0],[-1,-1,0,0],[0,0,-1,1],
    [0,0,-1,-1]]):
> N:=evalm(d1/sqrt(2)):
> d3:=array([[2,0,1,0],[0,2,0,1],[1,0,2,0],[0,1,0,2]]):
> m:=evalm(rh*d3*sqrt(2)/6):
> d5:=array([[1,0,-1,0],[0,0,0,0],[-1,0,1,0],[0,0,0,0]]):
> k:=evalm(EA/h*d5/sqrt(2)):
```

The mass matrix M3 now follows from (F6.9) and the stiffness matrix K3 from (F6.10):

```
> M3:=multiply(transpose(N),m,N);
```

$$M3 := \begin{bmatrix} 1/3\,rh\,2^{1/2} & 0 & 1/6\,rh\,2^{1/2} & 0 \\ 0 & 1/3\,rh\,2^{1/2} & 0 & 1/6\,rh\,2^{1/2} \\ 1/6\,rh\,2^{1/2} & 0 & 1/3\,rh\,2^{1/2} & 0 \\ 0 & 1/6\,rh\,2^{1/2} & 0 & 1/3\,rh\,2^{1/2} \end{bmatrix}$$

> K3:=multiply(transpose(N),k,N);

$$K3 := \begin{bmatrix} 1/4\frac{EA2^{1/2}}{h} & -1/4\frac{EA2^{1/2}}{h} & -1/4\frac{EA2^{1/2}}{h} & 1/4\frac{EA2^{1/2}}{h} \\ -1/4\frac{EA2^{1/2}}{h} & 1/4\frac{EA2^{1/2}}{h} & 1/4\frac{EA2^{1/2}}{h} & -1/4\frac{EA2^{1/2}}{h} \\ -1/4\frac{EA2^{1/2}}{h} & 1/4\frac{EA2^{1/2}}{h} & 1/4\frac{EA2^{1/2}}{h} & -1/4\frac{EA2^{1/2}}{h} \\ 1/4\frac{EA2^{1/2}}{h} & -1/4\frac{EA2^{1/2}}{h} & -1/4\frac{EA2^{1/2}}{h} & 1/4\frac{EA2^{1/2}}{h} \end{bmatrix}$$

Note that the change in $[M]$ is due to the difference in length only. We may conclude that for the element at hand the mass matrix does not depend on the orientation of the reference frame.

As the next example, derive the stiffness matrix K1 for the element 1 by transformation of that for the element 2, which implies $\alpha = -\pi/2$,

> N1:=array([[0,-1,0,0],[1,0,0,0],[0,0,0,-1],[0,0,1,0]]);

$$N1 := \begin{bmatrix} 0 & -1 & 0 & 0 \\ 1 & 0 & 0 & 0 \\ 0 & 0 & 0 & -1 \\ 0 & 0 & 1 & 0 \end{bmatrix}$$

> k1:=evalm(EA/h*d5):
> K1:=multiply(transpose(N1),k1,N1);

$$K1 := \begin{bmatrix} 0 & 0 & 0 & 0 \\ 0 & \frac{EA}{h} & 0 & -\frac{EA}{h} \\ 0 & 0 & 0 & 0 \\ 0 & -\frac{EA}{h} & 0 & \frac{EA}{h} \end{bmatrix}$$

F7. Generalizations

The one-dimensional elements considered so far are, of course, particular cases only. Fig. F5 shows some of the two- and three-dimensional elements, which are of use in the finite element analysis. As the triangular element displays, the nodes may be external and internal, with the node 7 being the internal one. Generally, the choice of shape, size and dimensionality of the element is a matter of engineering judgement, which depends on the type of structure at hand, considerations of the accuracy needed, etc. The following sections contain additional comments on this subject.

Next, the basic equation (F1.2) also needs a generalization. It must include the distributed force $\{p\}$, besides the nodal forces, and a possible residual stress. This equation takes the form

$$\{q\} + \{p\} = [k]\{w\} + \{p_0\} \tag{F7.1}$$

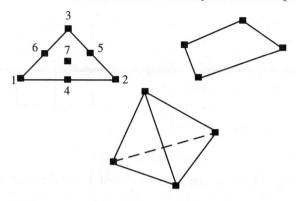

F5 Finite elements

where $\{p_0\}$ accounts for the residual stress. Each term in (F7.1) thus has a certain physical sense. If the intensity of the distributed force is so prescribed that it includes the nodal forces via delta-functions, the first term on the left-hand side may be omitted. The elemental equation (F7.1) takes the following form in a global coordinate system:

$$\{Q\} + \{P\} = [K]\{W\} + \{P_0\} \qquad (F7.2)$$

In the dynamic case this modifies to

$$\{Q\} + \{P\} = [M]\{W_{,tt}\} + [K]\{W\} + \{P_0\} \qquad (F7.3)$$

where the first term on the right-hand side is the force of inertia. In the following considerations we neglect the effect of residual stress, described by $\{P_0\}$.

Finally, the entire structure may also be considered as a finite element, as considerations of consistency show. This implies that (F7.2) and (F7.3) apply to the assembly of elements too. Nevertheless, modifications of notation for this case would be convenient. The external nodal forces applied to the structure will be denoted as $\{R\}$ and all relevant quantities labelled with the subscript 'a'. Then (F1.2) written for the assembly reads

$$\{R\}_a = [K]_a\{W\}_a \qquad (F7.4)$$

and (F7.3) reads

$$\{R\}_a + \{P\}_a = [M]_a\{W_{,tt}\}_a + [K]_a\{W\}_a + \{P_0\}_a \qquad (F7.5)$$

F8. Assembling

In the framework of preliminary considerations, assume that a structure or domain of interest has been represented by a mesh of finite elements, each of which has n degrees of freedom. If N is the number of degrees of freedom of the entire structure and Z the number of elements, then obviously

$$N < Zn \qquad (F8.1)$$

100 FINITE ELEMENT METHOD

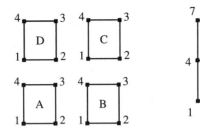

F6 Assembling elements

since the assembling of the elements would require the imposition of constraints.

Say, we assemble four elements as shown in Fig. F6, which means, among others, that the displacement at the common node must be the same. For example, the node is 3 for the element A, 4 for the element B, 1 for the element C and 2 for D. Assuming that the element has four degrees of freedom, we get 16 degrees of freedom for the set of four independent elements, but only nine of them for the assembly shown. Therefore, the nodal displacement vector for the assembly would contain nine components $\{W\}_a = [W_1, W_2, \ldots, W_9]^T$.

The displacement (or other field variable) should be matched not only at the nodes, but across the entire interelement boundary, which is of course relevant for two- and three-dimensional cases only. A proper choice of approximation for the field variable enables one to achieve this continuity. More precisely, there should be a correspondence between the number of terms in the approximation and the number of nodes.

Consider two triangular elements as shown in Fig. F7. Taking into account the six nodes available per element, a proper approximation for the field variable u may be set as

$$u = a_1 + a_2 x + a_3 y + a_4 x^2 + a_5 y^2 + a_6 xy \qquad \text{(F8.2)}$$

Therefore at the interface of two adjacent elements (see Fig. F7), the so-called nodal line (nodal surface), the field variable u behaves as a parabolic curve with three coefficients only. These, in turn, are governed by the nodal values u takes on at the three nodes of the nodal line, which ensures the desirable continuity.

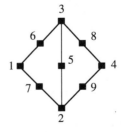

F7 To continuity across the nodal line

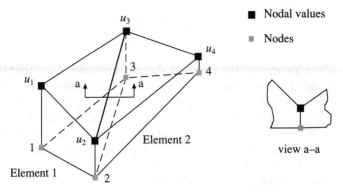

F8 To discontinuity of derivative

Note particularly that the continuity of the field variable does not imply the continuity of its derivatives across the nodal line. Fig. F8 shows two triangular elements and the associated nodal values of the field variable u. The view a–a exhibits the jump suffered by the derivative despite the continuity of the field u at the nodal line.

F9. Assembling via equilibrium equations

(MAPLE file vfem35)

Previous considerations have suggested a general form of equations governing the entire structure or the domain of interest. These are given by (F7.4) or in a more general form by (F7.5) and are reproduced below

$$\{R\}_a = [K]_a \{W\}_a \tag{F9.1}$$

$$\{R\}_a + \{P\}_a = [M]_a \{W_{,tt}\}_a + [K]_a \{W\}_a + \{P_0\}_a$$

Now one has to develop a procedure which would allow actual construction of the assembly matrices $[K]_a$ and $[M]_a$ from the elemental matrices. One of the ways to derive the stiffness matrix of the assembly is to construct the equilibrium equations, as the very form of (F9.1) assumes.

To this end, consider a typical node, j, which may be subjected to the external force $\mathbf{R}_j$. This force may also be thought of as the resultant of the external forces. On the other hand, each of the finite elements sharing the node j exerts the force $\mathbf{Q}_j$. The equilibrium equation for this node reads

102 FINITE ELEMENT METHOD

$$\mathbf{R}_j = \sum_{j=1}^{z} \mathbf{Q}_j \tag{F9.3}$$

where z is the number of the elements sharing the node. Note that (F9.1) and (F9.3) are in fact the same equations written in a different fashion. This similarity enables one to construct $[K]_a$ by considering (F9.3) for each of the nodes. We illustrate this procedure on the example of a truss shown in Fig. F9, which also indicates the external nodal forces and degrees of freedom. The latter are specified by the nodal displacement vector of the assembly $\{W\}_a = [W_1, W_2, \ldots, W_6]$.

As the figure displays, (F9.3) yields for the node 1

$$\begin{aligned} Q_1^{(1)} + Q_1^{(3)} &= R_1 \\ Q_2^{(1)} + Q_2^{(3)} &= R_2 \end{aligned} \tag{F9.4}$$

where the subscript denotes the degree of freedom and the superscript the element sharing the node. Similarly, for the node 2

$$\begin{aligned} Q_3^{(1)} + Q_3^{(2)} &= R_3 \\ Q_4^{(1)} + Q_4^{(2)} &= R_4 \end{aligned} \tag{F9.5}$$

and for the node 3

$$\begin{aligned} Q_5^{(2)} + Q_5^{(3)} &= R_5 \\ Q_6^{(2)} + Q_6^{(3)} &= R_6 \end{aligned} \tag{F9.6}$$

Expressing the left-hand sides of (F9.4), (F9.5) and (F9.6) in terms of the elemental equations:

$$\{Q\}_i = [K]_i \{W\}_i \qquad i = 1, 2, 3 \tag{F9.7}$$

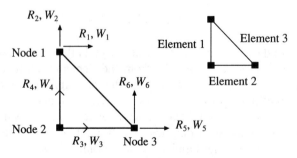

F9 Nodal forces and displacement

we get the relation (F9.1). Indeed, such an equation sets a direct dependence between $\{W\}_a$ and $\{R\}_a$. The matrix $[K]_a$ then follows from the command `genmatrix`.

Turning to computations, specify first the elemental matrices for a particular case setting $EA/h = 1$ (for convenience)

> K1:=array([[0,0,0,0],[0,1,0,-1],[0,0,0,0],[0,-1,0,1]]);

$$K1 := \begin{bmatrix} 0 & 0 & 0 & 0 \\ 0 & 1 & 0 & -1 \\ 0 & 0 & 0 & 0 \\ 0 & -1 & 0 & 1 \end{bmatrix}$$

> K2:=array([[1,0,-1,0],[0,0,0,0],[-1,0,1,0],[0,0,0,0]]);

$$K2 := \begin{bmatrix} 1 & 0 & -1 & 0 \\ 0 & 0 & 0 & 0 \\ -1 & 0 & 1 & 0 \\ 0 & 0 & 0 & 0 \end{bmatrix}$$

> d4:=array([[1,-1,-1,1],[-1,1,1,-1],[-1,1,1,-1],[1,-1,-1,1]]);

$$d4 := \begin{bmatrix} 1 & -1 & -1 & 1 \\ -1 & 1 & 1 & -1 \\ -1 & 1 & 1 & -1 \\ 1 & -1 & -1 & 1 \end{bmatrix}$$

> K3:=evalm(d4/(2*sqrt(2)));

$$K3 := \begin{bmatrix} 1/4\,2^{1/2} & -1/4\,2^{1/2} & -1/4\,2^{1/2} & 1/4\,2^{1/2} \\ -1/4\,2^{1/2} & 1/4\,2^{1/2} & 1/4\,2^{1/2} & -1/4\,2^{1/2} \\ -1/4\,2^{1/2} & 1/4\,2^{1/2} & 1/4\,2^{1/2} & -1/4\,2^{1/2} \\ 1/4\,2^{1/2} & -1/4\,2^{1/2} & -1/4\,2^{1/2} & 1/4\,2^{1/2} \end{bmatrix}$$

where results of Section F6 have been employed. Now state the elemental nodal displacement vectors $\text{Wd}i = \{W\}_i$, $i = 1, 2, 3$ in the gobal scheme:

> Wd1:=array([[W1],[W2],[W3],[W4]]);

$$Wd1 := \begin{bmatrix} W1 \\ W2 \\ W3 \\ W4 \end{bmatrix}$$

```
> Wd2:=array([[W3],[W4],[W5],[W6]]);
```

$$Wd2 := \begin{bmatrix} W3 \\ W4 \\ W5 \\ W6 \end{bmatrix}$$

```
> Wd3:=array([[W5],[W6],[W1],[W2]]);
```

$$Wd3 := \begin{bmatrix} W5 \\ W6 \\ W1 \\ W2 \end{bmatrix}$$

This enables one to state the elemental equilibrium equations (F9.7). Denoting $Qi = \{Q\}_i$, we get

```
> with(linalg,multiply,genmatrix,transpose):
> Q1:=multiply(K1,Wd1);
```

$$Q1 := \begin{bmatrix} 0 \\ W2 - W4 \\ 0 \\ -W2 + W4 \end{bmatrix}$$

```
> Q2:=multiply(K2,Wd2);
```

$$Q2 := \begin{bmatrix} W3 - W5 \\ 0 \\ -W3 + W5 \\ 0 \end{bmatrix}$$

```
> Q3:=multiply(K3,Wd3);
```

$$Q3 := \begin{bmatrix} 1/4\,2^{1/2}W5 - 1/4\,2^{1/2}W6 - 1/4\,2^{1/2}W1 + 1/4\,2^{1/2}W2 \\ -1/4\,2^{1/2}W5 + 1/4\,2^{1/2}W6 + 1/4\,2^{1/2}W1 - 1/4\,2^{1/2}W2 \\ -1/4\,2^{1/2}W5 + 1/4\,2^{1/2}W6 + 1/4\,2^{1/2}W1 - 1/4\,2^{1/2}W2 \\ 1/4\,2^{1/2}W5 - 1/4\,2^{1/2}W6 - 1/4\,2^{1/2}W1 + 1/4\,2^{1/2}W2 \end{bmatrix}$$

Proceeding with the equilibrium equations for the nodes, given by (F9.4), (F9.5) and (F9.6), we get

```
> R1:=Q1[1,1]+Q3[3,1];
```

$$R1 := -1/4\,2^{1/2}W5 + 1/4\,2^{1/2}W6 + 1/4\,2^{1/2}W1 - 1/4\,2^{1/2}W2$$

```
> R2:=Q1[2,1]+Q3[4,1];
```

$$R2 := W2 - W4 + 1/4\,2^{1/2}W5 - 1/4\,2^{1/2}W6 - 1/4\,2^{1/2}W1 + 1/4\,2^{1/2}W2$$

```
> R3:=Q1[3,1]+Q2[1,1];
    R3 := W3 - W5
> R4:=Q1[4,1]+Q2[2,1];
    R4 := - W2 + W4
> R5:=Q2[3,1]+Q3[1,1];
```
$$R5 := -W3 + W5 + 1/4\,2^{1/2}\,W5 - 1/4\,2^{1/2}\,W6 - 1/4\,2^{1/2}\,W1 + 1/4\,2^{1/2}\,W2$$
```
> R6:=Q2[4,1]+Q3[2,1];
```
$$R6 := -1/4\,2^{1/2}\,W5 + 1/4\,2^{1/2}\,W6 + 1/4\,2^{1/2}\,W1 - 1/4\,2^{1/2}\,W2$$

It is observed that the above six equations relate indeed between $\{R\}_a$ and $\{W\}_a$, as suggested by (F9.1). It remains to invoke genmatrix to derive $Ka = [K]_a$

```
> Ka:=genmatrix([R1,R2,R3,R4,R5,R6],[W1,W2,W3,W4,W5,W6]);
```

$$Ka := \begin{bmatrix} 1/4\,2^{1/2} & -1/4\,2^{1/2} & 0 & 0 & -1/4\,2^{1/2} & 1/4\,2^{1/2} \\ -1/4\,2^{1/2} & 1+1/4\,2^{1/2} & 0 & -1 & 1/4\,2^{1/2} & -1/4\,2^{1/2} \\ 0 & 0 & 1 & 0 & -1 & 0 \\ 0 & -1 & 0 & 1 & 0 & 0 \\ -1/4\,2^{1/2} & 1/4\,2^{1/2} & -1 & 0 & 1+1/4\,2^{1/2} & -1/4\,2^{1/2} \\ 1/4\,2^{1/2} & -1/4\,2^{1/2} & 0 & 0 & -1/4\,2^{1/2} & 1/4\,2^{1/2} \end{bmatrix}$$

where $EA/h = 1$.

F10. Assembling via connectivity

(MAPLE file vfem35)

The use of the equilibrium equations of the elements to derive the assembly stiffness matrix, demonstrated in the previous section, clarifies the underlying physical principle. A simpler and somewhat formal procedure for constructing the assembly matrices consists of the following steps: (i) if Z is the number of the elements and N the number of degrees of freedom of the assembly, set Z null matrices of size $N \times N$; (2) insert the entries of the elemental matrix $[K]_i$, $i = 1, 2, \ldots, Z$ into the ith null matrix, properly changing the elemental degrees of freedom to those of the assembly and constructing thereby the so-called ith expanded elemental matrix; (3) sum the expanded matrices to arrive at the assembly matrix.

106 FINITE ELEMENT METHOD

Consider again the truss problem of the previous section, setting for convenience $EA/h = 1$ and $N = 6$ and $Z = 3$. According to the above steps, set three 6×6 zero matrices Kexi, $i = 1, 2, 3$:

```
> Kex1:=array(1..6,1..6):
> Kex2:=array(1..6,1..6):
> Kex3:=array(1..6,1..6):
> for i to 6 do
> for j to 6 do
> Kex1[i,j]:=0:
> Kex2[i,j]:=0:
> Kex3[i,j]:=0:
> od:
> od:
```

(Note that, instead of the above loops, sparse may be used to specify zero matrices.)

The expanded elemental matrix $[K]_n^{exp}$, $n = 1, 2, 3$ follows from the zero matrix upon proper substitution from the corresponding elemental matrix in the global reference frame $[K]_n$, namely, the numbering of rows and columns must be changed to those of the assembly. We need therefore the correspondence law between the degrees of freedom of the element and those of the assembly. Denoting the rows and columns of $[K]_n$ by i and j, respectively, and the rows and columns of $[K]_n^{exp}$ by I and J, respectively, begin with the element 1. For convenience, we present again Fig. F9.

As the figure shows, the correspondence law for the element 1 is

$$i = I, j = J \quad \text{for} \quad i, j = 1, 2, 3 \tag{F10.1}$$

Here we have taken into account that in constructing $[k]_1$ from $[k]_2$ the rotation angle was prescribed as negative (see Section F6). Similarly, for the element 2, we get

$$i + 2 = I, j + 2 = J \quad \text{for} \quad i, j = 1, 2, 3 \tag{F10.2}$$

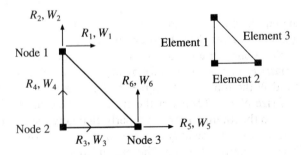

F9 Nodal forces and displacement

and for the element 3

$$i+4 = I, j+4 = J \quad \text{for} \quad i, j = 1, 2$$
$$i-2 = I, j-2 = J \quad \text{for} \quad i, j = 3, 4 \tag{F10.3}$$

In deriving (F10.3), we took into account that the angle of rotation of the element 3 was taken positive.

The transformation given by (F10.1), (F10.2) and (F10.3) specifies the *connectivity*, which maps the elemental numbering scheme to the assembly scheme and facilitates the programming, as shown below.

Begin the computation with the element 1. The transformation (F10.1) may be performed with the help of do (below Kex1 = $[K]_1^{\text{exp}}$)

```
> for i to 4 do
> for j to 4 do
> Kex1[i,j]:=K1[i,j]
> od:
> od:
> print(Kex1);
```

MATRIX(((0, 0, 0, 0, 0, 0], [0, 1, 0, −1, 0, 0], [0, 0, 0, 0, 0, 0], [0, −1, 0, 1, 0, 0], [0, 0, 0, 0, 0, 0], [0, 0, 0, 0, 0, 0]])

or in the prettyprint mode

```
> print(Kex1);
```

$$\begin{bmatrix} 0 & 0 & 0 & 0 & 0 & 0 \\ 0 & 1 & 0 & -1 & 0 & 0 \\ 0 & 0 & 0 & 0 & 0 & 0 \\ 0 & -1 & 0 & 1 & 0 & 0 \\ 0 & 0 & 0 & 0 & 0 & 0 \\ 0 & 0 & 0 & 0 & 0 & 0 \end{bmatrix}$$

Using (F10.2), we similarly derive Kex2 = $[K]_2^{\text{exp}}$

```
> for i to 4 do
> for j to 4 do
> Kex2[i+2,j+2]:=K2[i,j]
> od:
> od:
> print(Kex2);
```

$$\begin{bmatrix} 0 & 0 & 0 & 0 & 0 & 0 \\ 0 & 0 & 0 & 0 & 0 & 0 \\ 0 & 0 & 1 & 0 & -1 & 0 \\ 0 & 0 & 0 & 0 & 0 & 0 \\ 0 & 0 & -1 & 0 & 1 & 0 \\ 0 & 0 & 0 & 0 & 0 & 0 \end{bmatrix}$$

108 FINITE ELEMENT METHOD

As a simpler alternative to the do-statement, one may use a sequence to state the connectivity. For example, the connectivity for the element 3 given by (F10.3) may also be described by the sequence c3, as the following computation shows:

```
> c3:=5,6,1,2;
    c3 := 5, 6, 1, 2
> for i to 4 do for j to 4 do Kex3[c3[i],c3[j]]:=K3[i,j] od od:
> print(Kex3);
```

$$\begin{bmatrix} 1/4\,2^{1/2} & -1/4\,2^{1/2} & 0 & 0 & -1/4\,2^{1/2} & 1/4\,2^{1/2} \\ -1/4\,2^{1/2} & 1/4\,2^{1/2} & 0 & 0 & 1/4\,2^{1/2} & -1/4\,2^{1/2} \\ 0 & 0 & 0 & 0 & 0 & 0 \\ 0 & 0 & 0 & 0 & 0 & 0 \\ -1/4\,2^{1/2} & 1/4\,2^{1/2} & 0 & 0 & 1/4\,2^{1/2} & -1/4\,2^{1/2} \\ 1/4\,2^{1/2} & -1/4\,2^{1/2} & 0 & 0 & -1/4\,2^{1/2} & 1/4\,2^{1/2} \end{bmatrix}$$

Given the expanded matrices, we may complete the computation of the assembly matrix $[K]_a$ by

```
> Ka:=evalm(Kex1+Kex2+Kex3);
```

$$Ka := \begin{bmatrix} 1/4\,2^{1/2} & -1/4\,2^{1/2} & 0 & 0 & -1/4\,2^{1/2} & 1/4\,2^{1/2} \\ -1/4\,2^{1/2} & 1+1/4\,2^{1/2} & 0 & -1 & 1/4\,2^{1/2} & -1/4\,2^{1/2} \\ 0 & 0 & 1 & 0 & -1 & 0 \\ 0 & -1 & 0 & 1 & 0 & 0 \\ -1/4\,2^{1/2} & 1/4\,2^{1/2} & -1 & 0 & 1+1/4\,2^{1/2} & -1/4\,2^{1/2} \\ 1/4\,2^{1/2} & -1/4\,2^{1/2} & 0 & 0 & -1/4\,2^{1/2} & 1/4\,2^{1/2} \end{bmatrix}$$

Note that the result is identical to that of the previous section.

Instead of a sequence, the connectivity may be conveniently specified with the help of matrix c, whose entry c_{ij} is the global degree of freedom corresponding to the jth degree of freedom of the element i.

F11. Applications to a truss

(MAPLE file vfem35)

Consider the truss shown in Fig. F10, which has six degrees of freedom. Consequently, the assembly displacement vector is $\{W\}_a = [W_1, W_2, W_3, W_4, W_5, W_6]^T$. The supports of the truss indicate that

APPLICATIONS TO A TRUSS

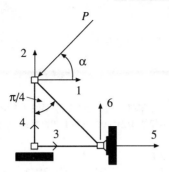

F10 Example of a truss

$$W_4 = W_5 = W_6 = 0 \tag{F11.1}$$

and the given external forces are

$$R_1 = -P\cos\alpha, \qquad R_2 = -P\sin\alpha, \qquad R_3 = 0 \tag{F11.2}$$

The equilibrium equations are

$$[K]_a\{W\}_a - \{R\}_a = 0 \tag{F11.3}$$

with $[K]_a$ specified in Section F10. Continuing that program, state `Wa` $= \{W\}_a$ and `Ra` $= \{R\}_a$

```
> Wa:=array([[W1],[W2],[W3],[W4],[W5],[W6]]);
```
  Wa := MATRIX(((W1],[W2],[W3],[W4],[W5],[W6]])

```
> Ra:=array([[R1],[R2],[R3],[R4],[R5],[R6]]);
```
  Ra := MATRIX(((R1],[R2],[R3],[R4],[R5],[R6]])

and then the left-hand side of (F11.3), denoted below as the matrix B:

```
> B:=evalm(multiply(Ka,Wa)-Ra);
```
  B := MATRIX(((− 1/4*2**(1/2)*W5 + 1/4*2**(1/2)*W6
  + 1/4*2**(1/2)*W1 − 1/4*2**(1/2)*W2 − R1], [W2 − W4
  + 1/4*2**(1/2)*W5 − 1/4*2**(1/2)*W6 − 1/4*2**(1/2)*W1
  + 1/4*2**(1/2)*W2 − R2], [W3 − W5 − R3], [− W2 + W4 − R4],
  [− W3 + W5 + 1/4*2**(1/2)*W5 − 1/4*2**(1/2)*W6 − 1/4*2**(1/2)*W1
  + 1/4*2**(1/2)*W2 − R5], [− 1/4*2**(1/2)*W5 + 1/4*2**(1/2)*W6
  + 1/4*2**(1/2)*W1 − 1/4*2**(1/2)*W2 − R6]])

Note that the matrix associated with this system of equations is singular, as it describes a 'floating' truss. The first step therefore is to incorporate the conditions given by (F11.1) and (F11.2), using `subs`. It may be convenient to do this with the help of array `s[i]`, whose elements are the equations stated earlier in B:

110 FINITE ELEMENT METHOD

```
> for j to 6 do
> s[j]:=subs({R1= -P*cos(alpha),R2= -P*sin(alpha),R3=0,
  W4=0,W5=0,W6=0},
> B[j,1]);
> od;
```

$$s[1] := 1/4*2**(1/2)*W1 - 1/4*2**(1/2)*W2 + P*cos(alpha)$$
$$s[2] := W2 - 1/4*2**(1/2)*W1 + 1/4*2**(1/2)*W2 + P*sin(alpha)$$
$$s[3] := W3$$
$$s[4] := -W2 - R4$$
$$s[5] := -W3 - 1/4*2**(1/2)*W1 + 1/4*2**(1/2)*W2 - R5$$
$$s[6] := 1/4*2**(1/2)*W1 - 1/4*2**(1/2)*W2 - R6$$

It remains to solve the system of equations for the unknown quantities:

```
> d30:=solve({s[m] $m=1..6},{W1,W2,W3,R4,R5,R6});
```

d30 := {W3 = 0, R4 = P*cos(alpha) + P*sin(alpha),
W1 = − P*cos(alpha) − P*sin(alpha) − 2*2**(1/2)*P*cos(alpha),
R6 = − P*cos(alpha), R5 = P*cos(alpha), W2 = − P*cos(alpha) −
P*sin(alpha)}

The next example deals with the truss shown in Fig. F11.

Considering each of the bars as the finite element, one may derive the assembly stiffness matrix in the way demonstrated earlier.

A more interesting approach may take advantage of the symmetry and view the truss as the assembly of two elements shown in Fig. F12, for which [K] has been already derived in Section F10. However, by this method the diagonal bar would, figuratively speaking, appear twice. This means that the use of the above [K]-matrices is permissible if this bar has double the stiffness of the others. Otherwise, the matrices must be recomputed.

It is seen that for the element 1 the stiffness matrix $[K] = [k]$ is given by the expression Ka of the previous section, but for the element 2 the rotation to the angle $\alpha = \pi$ is necessary. Using (F6.11) and taking into account that the element has six degrees of freedom, we get first the rotation matrix N2

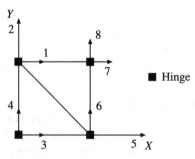

F11 Example of a truss

APPLICATIONS TO A TRUSS

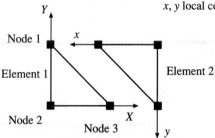

F12 Substructuring

```
> array(1..6,1..6,identity);
```

MATRIX(((1, 0, 0, 0, 0, 0], [0, 1, 0, 0, 0, 0], [0, 0, 1, 0, 0, 0],
[0, 0, 0, 1, 0, 0], [0, 0, 0, 0, 1, 0], [0, 0, 0, 0, 0, 1]])

```
> N2:=evalm(-1*");
```

N2 := MATRIX(((−1, 0, 0, 0, 0, 0], [0, −1, 0, 0, 0, 0], [0, 0, −1, 0, 0, 0],
[0, 0, 0, −1, 0, 0], [0, 0, 0, 0, −1, 0], [0, 0, 0, 0, 0, −1]])

The stiffness matrix K2 follows from (F6.10)

```
> k2:=Ka:
> gc();
> K2:=multiply(transpose(N2),k2,N2);
```

K2 := MATRIX(((1/4*2**(1/2), −1/4*2**(1/2), 0, 0, −1/4*2**(1/2),
1/4*2**(1/2)], [−1/4*2**(1/2), 1 + 1/4*2**(1/2), 0, −1, 1/4*2**(1/2),
−1/4*2**(1/2)], [0, 0, 1, 0, −1, 0], [0, −1, 0, 1, 0, 0], [−1/4*2**(1/2),
1/4*2**(1/2), −1, 0, 1 + 1/4*2**(1/2), −1/4*2**(1/2)], [1/4*2**(1/2),
−1/4*2**(1/2), 0, 0, −1/4*2**(1/2), 1/4*2**(1/2)]])

The next step is to construct the expanded matrices, which should be of size 8 × 8:

```
> Kex1:=array(1..8,1..8,[]):
> Kex2:=array(1..8,1..8,[]):
> for i to 8 do
> for j to 8 do
> Kex1[i,j]:=0:
> Kex2[i,j]:=0:
> od:
> od:
```

112 FINITE ELEMENT METHOD

For the element 1 no change of numbering is needed:

```
> for i to 6 do
> for j to 6 do
> Kex1[i,j]:=Ka[i,j]
> od:
> od:
> print(Kex1);
```

MATRIX(((1/4*2**(1/2), −1/4*2**(1/2), 0, 0, −1/4*2**(1/2), 1/4*2**(1/2), 0, 0],
[−1/4*2**(1/2), 1 + 1/4*2**(1/2), 0, −1, 1/4*2**(1/2),
−1/4*2**(1/2), 0, 0], [0, 0, 1, 0, −1, 0, 0, 0], [0, −1, 0, 1, 0, 0, 0, 0],
[−1/4*2**(1/2), 1/4*2**(1/2), −1, 0, 1 + 1/4*2**(1/2), −1/4*2**(1/2), 0, 0],
[1/4*2**(1/2), −1/4*2**(1/2), 0, 0, −1/4*2**(1/2), 1/4*2**(1/2), 0, 0],
[0, 0, 0, 0, 0, 0, 0, 0], [0, 0, 0, 0, 0, 0, 0, 0]])

For the element 2, in view of the above rotation, the connectivity is given by c2 below:

```
> c2:=5,6,7,8,1,2;
```

c2 := 5, 6, 7, 8, 1, 2

which leads to the following expanded stiffness matrix for the element 2:

```
> for i to 6 do for j to 6 do Kex2[c2[i],c2[j]]:=K2[i,j] od od:
> print(Kex2);
```

MATRIX(((1 + 1/4*2**(1/2), −1/4*2**(1/2), 0, 0, −1/4*2**(1/2),
1/4*2**(1/2), −1,0], [−1/4*2**(1/2), 1/4*2**(1/2), 0, 0, 1/4*2**(1/2),
−1/4*2**(1/2), 0, 0], [0, 0, 0, 0, 0, 0, 0, 0], [0, 0, 0, 0, 0, 0, 0, 0],
[−1/4*2**(1/2), 1/4*2**(1/2), 0, 0, 1/4*2**(1/2), −1/4*2**(1/2), 0, 0],
[1/4*2**(1/2), −1/4*2**(1/2), 0, 0, −1/4*2**(1/2), 1 + 1/4*2**(1/2), 0, −1],
[−1, 0, 0, 0, 0, 0, 1, 0], [0, 0, 0, 0, 0, 0, −1, 0, 1]])

The stiffness matrix of the assembly is

```
> Ka:=evalm(Kex1+Kex2);
```

Ka := MATRIX(((1/2*2**(1/2) + 1, −1/2*2**(1/2), 0, 0, −1/2*2**(1/2),
1/2*2**(1/2), −1, 0], [−1/2*2**(1/2), 1/2*2**(1/2) + 1, 0, −1, 1/2*2**(1/2),
−1/2*2**(1/2), 0, 0], [0, 0, 1, 0, −1, 0, 0, 0], [0, −1, 0, 1, 0, 0, 0, 0],
[−1/2*2**(1/2), 1/2*2**(1/2), −1, 0, 1/2*2**(1/2) + 1, −1/2*2**(1/2), 0, 0],
[1/2*2**(1/2), −1/2*2**(1/2), 0, 0, −1/2*2**(1/2), 1/2*2**(1/2) + 1, 0, −1],
[−1, 0, 0, 0, 0, 0, 1, 0],[0, 0, 0, 0, 0, −1, 0, 1]])

Now consider the case shown in Fig. F13.

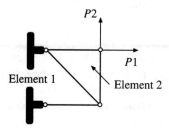

F13 Example of a truss

Stating the assembly displacement matrix Wa and external force matrix Ra,

```
> Wa:=array(1..8,1..1,[]):
> Ra:=array(1..8,1..1,[]):
> for i to 8 do
> Wa[i,1]:=W.i:
> Ra[i,1]:=R.i:
> od:
```

one may formulate the equilibrium equations:

```
> B:=evalm(multiply(Ka,Wa)-Ra);
```

$B :=$ MATRIX(((1/2*2**(1/2)*W1 + W1−1/2*2**(1/2)*W2−1/2*2**(1/2)*W5 + 1/2*2**(1/2)*W6−W7−R1], [−1/2*2**(1/2)*W1 + 1/2*2**(1/2)*W2 + W2−W4 + 1/2*2**(1/2)*W5−1/2*2**(1/2)*W6−R2], [W3−W5−R3], [−W2 + W4−R4], [−1/2*2**(1/2)*W1 + 1/2*2**(1/2)*W2−W3 + 1/2*2**(1/2)*W5 + W5−1/2*2**(1/2)*W6−R5], [1/2*2**(1/2)*W1−1/2*2**(1/2)*W2−1/2*2**(1/2)*W5 + 1/2*2**(1/2)*W6 + W6−W8−R6], [−W1 + W7−R7], [−W6 + W8−R8]])

Introducing the support conditions and the given external forces (see Fig. F11 for the degrees of freedom), we get the solution by

```
> for i to 8 do
> s[i]:=subs({W1=0,W2=0,W3=0,W4=0,R5=0,R6=0,R7=P1,R8=P2},
  B[i,1]);
> od:
> solve({s[G] $G=1..8},{R1,R2,R3,R4,W5,W6,W7,W8});
```

{R3 = −P2, W5 = P2, R4 = 0, W7 = P1, W8 = 2**(1/2)*P2 + 2*P2,
R1 = P2−P1, R2 = −P2, W6 = 2**(1/2)*P2 + P2}

Note again that these results have been obtained for $AE/h = 1$ for all the elements. The values of the displacement must therefore be multiplied by h/AE.

114 FINITE ELEMENT METHOD

F12. Inhomogeneous beam

(MAPLE file vfem36)

Consider a composite beam shown in Fig. F14. It consists of two elements with the stiffnesses EI_1 and EI_2, respectively, and the lengths h_1 and h_2, respectively.

These elements are firmly bonded together at the interface. There are six degrees of freedom, associated with the three nodes, which gives rise to the assembly displacement matrix

$$\{W\}_a = [W_1, W_2, \ldots, W_6]^T \qquad (F12.1)$$

and the assembly force matrix

$$\{R\}_a = [R_1, R_2, \ldots, R_6]^T \qquad (F12.2)$$

These are related by

$$\{R\}_a = [K]_a \{W\}_a \qquad (F12.3)$$

where $[K]_a$ is the assembly stiffness matrix.

The first step is to construct the elemental stiffness matrix $[k]$ in the local coordinate system. This has been derived in Section F1. So, just copy that program:

```
> w:=sum(a[i]*x^i,i=0..3):
> wx:=diff(w,x):
> d4:={w1=subs(x=0,w),w3=subs(x=h,w),f2=subs(x=0,wx),
   f4=subs(x=h,wx)}:
> solution:=solve(d4,{a[0],a[1],a[2],a[3]}):
> subs(",w):
> w:=":
> subs(solution,wx):
> wx:=":
> wxx:=diff(wx,x):
> wxxx:=diff(",x):
> v[1]:=subs(x=0,wxxx):
> v[2]:=-subs(x=0,wxx):
```

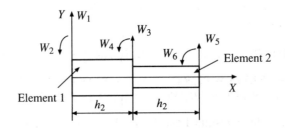

F14 Alternating beam elements

```
> v[3]:=-subs(x=h,wxxx):
> v[4]:=subs(x=h,wxx):
> d10:=[v[1],v[2],v[3],v[4]]:
> with(linalg,genmatrix):
> kk:=genmatrix(d10, [w1,f2,w3,f4]):
> k:=evalm(EI*kk):
> print(k);
```

MATRIX(((12*EI/h**3, 6*EI/h**2, −12*EI/h**3, 6*EI/h**2],
[6*EI/h**2, 4*EI/h,−6*EI/h**2, 2*EI/h], [-12*EI/h**3, −6*EI/h**2,
12*EI/h**3, −6*EI/h**2], [6*EI/h**2, 2*EI/h, −6*EI/h**2, 4*EI/h]])

Since no rotation of the matrices is needed, we get the 'global' matrices K1 and K2 by a mere substitution of the appropriate stiffnesses for *EI* in k:

```
> for i to 4 do
> for j to 4 do
> K1[i,j]:=subs({h=h1,EI=E1},k[i,j]):
> K2[i,j]:=subs({h=h2,EI=E2},k[i,j]):
> od:
> od:
```

where for brevity we set $EI_1 =$ E1 and $EI_2 =$ E2. Then begin to construct the two expanded matrices by formulating zero matrices:

```
> Kex1:=array(sparse,1..6,1..6,[]):
> Kex2:=array(sparse,1..6,1..6,[]):
```

Changing properly the degrees of freedom for the element 2 according to Fig. F14, we complete the expanded matrices:

```
> for i to 4 do
> for j to 4 do
> Kex1[i,j]:=K1[i,j]:
> Kex2[i+2,j+2]:=K2[i,j]
> od:od:
```

It remains to sum them up to arrive at the assembly stiffness matrix:

```
>Ka:=evalm(Kex1+Kex2);
```

Ka := MATRIX(((12*E1/h1**3, 6*E1/h1**2, −12*E1/h1**3,
6*E1/h1**2, 0, 0], [6*E1/h1**2, 4*E1/h1, −6*E1/h1**2, 2*E1/h1, 0, 0],
[−12*E1/h1**3, −6*E1/h1**2, 12*E1/h1**3 + 12*E2/h2**3,
−6*E1/h1**2 + 6*E2/h2**2, −12*E2/h2**3, 6*E2/h2**2], [6*E1/h1**2,
2*E1/h1, −6*E1/h1**2 + 6*E2/h2**2, 4*E1/h1 + 4*E2/h2, −6*E2/h2**2,
2*E2/h2], [0, 0, −12*E2/h2**3, −6*E2/h2**2, 12*E2/h2**3,
−6*E2/h2**2], [0, 0, 6*E2/h2**2,2*E2/h2, −6*E2/h2**2, 4*E2/h2]])

116 FINITE ELEMENT METHOD

F13. Applications to a beam

(MAPLE file vfem36)

Consider first an inhomogeneous cantilever subjected to a force P at the end, as shown in Fig. F15. As in the previous section, formulate the equilibrium equations in their general form:

```
> Wa:=array(1..6,1..1):
> Ra:=array(1..6,1..1):
> for i to 6 do
> Wa[i,1]:=W.i:
> Ra[i,1]:=R.i
> od:
> with(linalg,multiply);
```

   [multiply]

```
> B:=evalm(multiply(Ka,Wa)-Ra):
```

Then introduce the s-array for the above B-equations and substitute the boundary conditions. Note that it is not necessary to indicate explicitly the sign of the quantities imposed at the nodes, as the solution has an analytical form:

```
> s:=array(1..6):
> for i to 6 do
> s[i]:=subs({W1=0,W2=0,R3=0,R4=0,R5=-P,R6=0},B[i,1])
> od:
```

It remains to solve the above equations for the unknown quantities:

```
> convert(s,set):
> solve(", {R1,R2,W3,W4,W5,W6});
```

   $\{W6 = -1/2*P*(E2*h1**2 + 2*h2*E2*h1 + E1*h2**2)/E2/E1,\ R1 = P,$
   $W5 = -1/3/E2/E1*P*(3*h1**2*h2*E2 + 3*h2**2*E2*h1 + E1*h2**3 +$
   $E2*h1**3),\ R2 = P*(h1+h2),\ W4 = -1/2*P*h1*(2*h2+h1)/E1,$
   $W3 = -1/6*P*h1**2*(3*h2+2*h1)/E1\}$

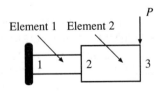

F15 Example of a beam

APPLICATIONS TO A BEAM 117

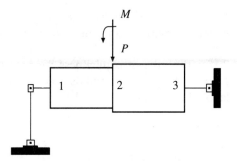

F16 Example of a beam

The next example is a hinged beam, as shown in Fig. F16.

We state the boundary conditions in the c-set given below, substitute this in the governing equations B and solve for the unknown quantities,

```
> c:={W1=0,R2=0,W5=0,R6=0,R3=-P,R4=-M}:
> for i to 6 do
> s[i]:=subs(c,B[i,1]):
> od:
> convert(s,set):
> solve(",{R1,W2,R5,W6,W3,W4}):map(factor,");
```

{W3 = − 1/3*h2*h1*(− E2*M*h1**2 + E2*h1**2*P*h2 + h1* h2**2*
E1*P + h2**2*E1*M)/E2/E1/(h1 + h2)**2, R5 = (P*h1 + M)/(h1 + h2),
W6 = 1/6*(h2**3*E1*h1* P + E1*h2**3*M + 3*h1**2*h2**2*
E1*P + 3*h1*h2**2*E1*M − 2*E2*h1**3*M + 2*E2*h1**3*P*h2)/E2/E1/
(h1 + h2)**2, R1 = (− M + P*h2)/(h1 + h2), W2 = − 1/6/E2*(2*
h2**3*E1*h1* P + 2*E1*h2**3*M − E2*h1**3*M + E2*h1**3*P*h2 −
3*E2*h2*M*h1**2 + 3*E2*h1**2*P*h2**2)/E1/(h1 + h2)**2,
W4 = 1/3*(− h2**3*E1*h1*P − E1*h2**3*M − E2*h1**3*M
+ E2*h1**3*P*h2)/E2/E1/(h1 + h2)**2}

It is left to the reader to verify whether these results reproduce those of the homogeneous beam as a particular case.

F14. Automatic generation of an assembly matrix

(MAPLE file vfem37)

The previous program may be further improved in a way which would allow for automatic derivation of the assembly stiffness matrix for an arbitrary number of elements. Such a program is given below. It includes, as earlier, the following steps: (1) formulation of the approximating polynomial;

118 FINITE ELEMENT METHOD

(2) formulation of the shape functions and the elemental matrix; (3) formulation of the expanded matrices and then the assembly matrix.

Below Z is the number of elements and N the number of degrees of freedom. Though $N = 2Z + 2$, for the sake of explicitness we prescribe both of these values. As indicated earlier, the first commands specify the stiffness matrix of the single element denoted as kf:

```
> w:=sum(a[i]*x^i,i=0..3):
> wx:=diff(w,x):
> {ww[1]-subs(x=0,w),ww[2]-subs(x=0,wx),ww[3]-subs
   (x=h,w),
> ww[4]-subs(x=h,wx)}:
> d4:=solve(",{a[0],a[1],a[2],a[3]}):
> d5:=subs(",w):
> for i to 4 do
> d[i](x):=coeff(collect(d5,ww[i]),ww[i]):
> od:
> kf:=array(1..4,1..4):
> for i to 4 do
> for j to 4 do
> if i<=j then
> kf[i,j]:=EI*int(diff(d[i](x),x$2)*diff(d[j](x),x$2),
   x=0..h):
> else kf[i,j]:=kf[j,i] fi:
> od:
> od:
```

Next, define the procedure Kautomat(Z,N), which would derive the assembly matrix. Its first step is the statement of the Z elemental matrices:

```
> i:='i':j:='j':
> Kautomat:=proc(Z,N)
> for q to Z do kff.q:=array(1..4,1..4):
> for i to 4 do for j to 4 do
> kff.q[i,j]:=subs({EI=EI.q,h=h.q},kf[i,j]):
> od:
> od:
```

where EI.q and h.q are the stiffness and length of the q-element.

Going over to the expanded matrices, state the connectivity for the q-element:

$$i = I + 2(q-1)$$
$$j = J + 2(q-1), \qquad q = 1, 2, \ldots, Z \qquad \text{(F14.1)}$$

AUTOMATIC GENERATION OF AN ASSEMBLY MATRIX

which is of use below:

```
> i:='i':j:='j':
> ke.q:=array(1..N,1..N,sparse):
> for i to 4 do for j to 4 do ke.q[i+(q-1)*2,
   j+(q-1)*2]:=kff.q[i,j];
> od:
> od:
> od:
```

It remains to sum the expanded matrices to arrive at the assembly matrix Ka:

```
> i:='i':j:='j':
> Ka:=array(1..N,1..N):
> for i to N do for j to N do
> Ka[i,j]:=sum('ke.q[i,j]','q'=1..Z):
> od:
> od:
> print(Ka);
> end:
```

Note that for the given Z this program provides an analytical representation of $[K]_a =$ Ka, which may be used for many purposes, for example, optimization. The following section presents a proper example of this application.

We save the above program as a separate file autobeam.m and consider a clamped composite beam under the force P_0 applied at $x = e$ (Fig. F17).

In agreement with the discussion of Section F2, this force may be represented by the equivalent nodal forces applied at the nodes 2 and 3. These forces follow from (F2.10) as

$$\{p\} = P_0 \int_0^h \delta(x-e)\{d(x)\} \, dx \qquad (F14.2)$$

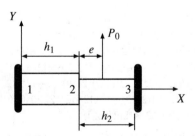

F17 Composite beam

120 FINITE ELEMENT METHOD

where $\{d(x)\}$ is the matrix of interpolation functions. This gives

$$\{p\}_2 = [d_1(e)\, d_2(e)\, d_3(e)\, d_4(e)]^T \qquad (F14.3)$$

in the local scheme and

$$\{P\}_2 = [0\, 0\, d_1(e)\, d_2(e)\, d_3(e)\, d_4(e)]^T \qquad (F14.4)$$

in the global scheme. Here use has been made of the connectivity (F14.1) for the element 2. Setting for the sake of certainty $e = h_2/2$, state first (F14.3) and (F14.4):

```
> p:=array(1..4,sparse):
> for i to 4 do
> p[i]:=subs({x=h2/2,h=h2},d[i](x)*P0):
> od:
> print(p);
```

   VECTOR([1/2*P0, 1/8*h2*P0, 1/2*P0, −1/8*h2*P0])

```
> Pa:=array(1..6,1..1,sparse):
> for i to 4 do
> Pa[i+2,1]:=p[i]:
> od:
> print(Pa);
```

   MATRIX(((0], [0], [1/2*P0], [1/8*h2*P0], [1/2*P0], [−1/8*h2*P0]])

then specify the nodal quantities:

```
> Wa:=array(1..6,1..1):
> Ra:=array(1..6,1..1):
> for i to 6 do
> Wa[i,1]:=W.i;
> Ra[i,1]:=R.i;
> od:
```

and the assembly stiffness matrix:

```
> Kautomat(2,6);
```

   MATRIX(((12*EI1/h1**3, 6*EI1/h1**2, −[12*EI1/h1**3, 6*EI1/h1**2, 0, 0],
   [6*EI1/h1**2, 4*EI1/h1, − 6*EI1/h1**2, 2*EI1/h1, 0, 0],
   [−12*EI1/h1**3, − 6*EI1/h1**2,12*EI1/h1**3 + 12*EI2/h2**3,
   − 6*EI1/h1**2 + 6*EI2/h2**2, −12*EI2/h2**3, 6*EI2/h2**2],
   [6*EI1/h1**2, 2*EI1/h1, − 6*EI1/h1**2 + 6*EI2/h2**2, 4*EI1/h1
   + 4*EI2/h2, −6*EI2/h2**2, 2*EI2/h2], [0, 0, −12*EI2/h2**3,
   −6*EI2/h2**2, 12*EI2/h2**3, − 6*EI2/h2**2], [0, 0, 6*EI2/h2**2,
   2*EI2/h2, −6*EI2/h2**2, 4*EI2/h2]])

Now we are in a position to state the equilibrium equations and substitute the given boundary values:

```
> with(linalg,multiply):
> B:=evalm(multiply(Ka,Wa)-Ra-Pa):
> s:=array(1..6):
> for i to 6 do
> s[i]:=subs({W1=0,W2=0,R3=0,R4=0,W5=0,W6=0},B[i,1]):
> od;
```

$$s[1] := 6*EI1*(-2*W3 + W4*h1)/h1**3 - R1$$
$$s[2] := 2*EI1*(-3*W3 + W4*h1)/h1**2 - R2$$
$$s[3] := -6*(-2*W3*EI1*h2**3 - 2*W3*EI2*h1**3$$
$$+ W4*h1*h2**3*EI1 - W4*h1**3*h2*EI2)/h1**3/h2**3 - 1/2*P0$$
$$s[4] := 2*(-3*W3*EI1*h2**2 + 3*W3*EI2*h1**2$$
$$+ 2*W4*h1*h2**2*EI1 + 2*W4*h1**2*h2*EI2)/h1**2/h2**2 - 1/8*h2*P0$$
$$s[5] := -6*EI2*(2*W3 + W4*h2)/h2**3 - R5 - 1/2*P0$$
$$s[6] := 2*EI2*(3*W3 + W4*h2)/h2**2 - R6 + 1/8*h2*P0$$

Solving for the unknowns, we get

```
> convert(s,set):
> sol:=solve(",{R1,R2,W3,W4,R5,R6}):
```

where the display of lengthy results has been prevented. Nevertheless, verify the solution for a particular case of the uniform beam:

```
> partcase:=subs({EI2=EI1,h2=h,h1=h},sol):
> simplify(partcase);
```

$$\{R1 = -5/32*P0, W3 = 1/48*h**3*P0/EI1, W4 = 1/64*h**2*P0/EI1,$$
$$R5 = -27/32*P0, R6 = 9/32*h*P0, R2 = -3/32*h*P0\}$$

F15. Optimization

(MAPLE file vfem37)

As a next example of the usefulness of the above analytical results, consider a hinged beam, which consists of the alternating elements with the same length h and the stiffness EI_1 and EI_2, respectively (Fig. F18). The beam is subject to a force P_0.

The overall stiffness of the structure K is given by

$$K = P_0/W_0 \qquad (F15.1)$$

where W_0 is the deflection of the application point, and is a function of the above EI_1 and EI_2. Denoting

$$a = EI_2/EI_1 \qquad (F15.2)$$

122 FINITE ELEMENT METHOD

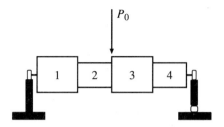

F18 Optimizing the beam

we may set $K = K(a)$. The subject of interest here is that value of a which would maximize F. To this end, we first solve the problem in general terms with the help of the program of Section F14.

Noting that the number of the elements $Z = 4$ and the number of degrees of freedom $N = 10$, we state the nodal quantities

```
> Wa:=array(1..10,1..1):Ra:=copy("):
> for i to 10 do
> Wa[i,1]:=W.i:Ra[i,1]:=R.i:
> od:
```

and then specify the assembly stiffness matrix

```
> Kautomat(4,10):
```

MATRIX(((12*EI1/h1**3, 6*EI1/h1**2, −12*EI1/h1**3, 6*EI1/h1**2, 0, 0, 0, 0, 0,0], [6*EI1/h1**2, 4*EI1/h1, −6*EI1/h1**2, 2*EI1/h1, 0, 0, 0, 0, 0, 0], [−12*EI1/h1**3, −6*EI1/h1**2, 12*EI1/h1**3 + 12*EI2/h2**3, −6*EI1/h1**2 + 6*EI2/h2**2, −12*EI2/h2**3, 6*EI2/h2**2, 0, 0, 0, 0], [6*EI1/h1**2, 2*EI1/h1, −6*EI1/h1**2 + 6*EI2/h2**2, 4*EI1/h1 + 4*EI2/h2, −6*EI2/h2**2, 2*EI2/h2, 0, 0, 0, 0], [0, 0, −12*EI2/h2**3, −6*EI2/h2**2, 12*EI2/h2**3 + 12*EI3/h3**3, −6*EI2/h2**2 + 6*EI3/h3**2, −12*EI3/h3**3, 6*EI3/h3**2, 0, 0], [0, 0, 6*EI2/h2**2, 2*EI2/h2, −6*EI2/h2**2 + 6*EI3/h3**2, 4*EI2/h2 + 4*EI3/h3, −6*EI3/h3**2, 2*EI3/h3, 0, 0], [0, 0, 0, 0, −12*EI3/h3**3, −6*EI3/h3**2, 12*EI3/h3**3 + 12*EI4/h4**3, −6*EI3/h3**2 + 6*EI4/h4**2, −12*EI4/h4**3, 6*EI4/h4**2], [0, 0, 0, 0, 6*EI3/h3**2, 2*EI3/h3, −6*EI3/h3**2 + 6*EI4/h4**2, 4*EI3/h3 + 4*EI4/h4, −6*EI4/h4**2, 2*EI4/h4], [0, 0, 0, 0, 0, 0, −12*EI4/h4**3, −6*EI4/h4**2, 12*EI4/h4**3, −6*EI4/h4**2], [0, 0, 0, 0, 0, 0, 6*EI4/h4**2, 2*EI4/h4, −6*EI4/h4**2, 4*EI4/h4]])

Now formulate the equilibrium equations, substitute the given nodal values and solve for the unknown quantities:

```
> B:=evalm(multiply(Ka,Wa)-Ra):
```

```
> s:=array(1..10):
> for i to 10 do
> s[i]:=subs({W1=0,R2=0,W9=0,R10=0,
> R3=0,R4=0,R5=-P,R6=0,R7=0,R8=0},
> B[i,1]):
> od:
> convert(s,set):
> sol1:=solve(",{R1,W2,R9,W10,W3,W4,W5,W6,W7,W8}):
> assign(sol1):
```

where the display of lengthy expressions has been prevented.

Now we have at our disposal the analytical results for the previously unknown quantities. The quantity of interest is W5 = W_0. Substitute the length and stiffness of the elements and specify W5:

```
> subs({h1=h,h2=h,h3=h,h4=h,EI1=EI,EI2=a*EI,EI3=EI,
  EI4=a*EI},W5):
> W5:=simplify(");
```

$$W5 := -2/3*P*h**3/a/EI*(1+a)$$

Therefore the overall stiffness K is

$$K = \frac{3aEI}{2(a+1)h^3} \qquad (F15.3)$$

which enables one to study the limit cases. For the rigid elements 1 and 3, we get

$$EI_1 = EI \to \infty$$
$$a \to 0 \qquad (F15.4)$$

and (F15.3) yields

$$K = \frac{3EI_2}{2h^3} \qquad (F15.5)$$

where use has been made of (F15.2). Conversely, if the elements 2 and 4 become rigid, then

$$EI_2 = aEI \to \infty$$
$$a \to \infty \qquad (F15.6)$$

and (F15.3) similarly yields

$$K = \frac{3EI_1}{2h^3} \qquad (F15.7)$$

F16. Imposition of constraints

(MAPLE file vfem38)

So far the boundary (support) conditions have been substituted in a straightforward way in the equilibrium equations and the command `solve` provided a solution. With the increased number of equations to be solved simultaneously this procedure may become difficult to perform, so any reduction in this number is useful. One of the methods of achieving this is the elimination of rows and columns associated with the vanishing nodal displacement. At the same time, this allows for a convenient incorporation of a given constraint.

Consider for simplicity the case of a 3×3 assembly matrix. Generate first $[K]_a =$ Ka, $\{W\}_a =$ Wa and $\{R\}_a =$ Ra and then the equilibrium equations in their general forms:

```
> Ka:=array(1..3,1..3):
> Wa:=array(1..3,1..1):
> Ra:=array(1..3,1..1):
> for i to 3 do
> for j to 3 do
> Ka[i,j]:=k.i.j;
> od;
> od;
> for i to 3 do
> Wa[i,1]:=W.i;
> Ra[i,1]:=R.i;
> od:
> with(linalg,multiply):
> d3:=evalm(multiply(Ka,Wa)-Ra);
```

$$d3 := \begin{bmatrix} k11 \; W1 + k12 \; W2 + k13 \; W3 - R1 \\ k21 \; W1 + k22 \; W2 + k23 \; W3 - R2 \\ k31 \; W1 + k32 \; W2 + k33 \; W3 - R3 \end{bmatrix}$$

Assume that the boundary conditions impose $W_2 =$ W2 $= 0$ and substitute this in d3

```
> d4:=subs(W2=0,");
```

$$d4 := \begin{bmatrix} k11 & W1+k13 & W3-R1 \\ k21 & W1+k23 & W3-R2 \\ k31 & W1+k33 & W3-R3 \end{bmatrix}$$

Note that it is possible to resolve the first and third equations in d4 independently of the second one and that the matrix associated with the first and third equations follows by eliminating the second row and second column from $[K]_a$ and the second row from $\{W\}_a$ and $\{R\}_a$. Quantities obtained in such a way are referred to as *reduced* quantities. The reduced equilibrium equations are

$$[K]_r \{W\}_r - \{R\}_r = 0 \tag{F16.1}$$

The convenient command to construct the reduced quantities $\text{Kr} = [K]_r$, $\text{Wr} = \{W\}_r$ and $\text{Rr} = \{R\}_r$ is submatrix:

> with(linalg,submatrix):
> Kr:=submatrix(Ka,[1,3],[1,3]);

$$Kr := \begin{bmatrix} k11 & k13 \\ k31 & k33 \end{bmatrix}$$

> Wr:=submatrix(Wa,[1,3],1..1);

$$Wr := \begin{bmatrix} W1 \\ W3 \end{bmatrix}$$

> Rr:=submatrix(Ra,[1,3],1..1);

$$Rr := \begin{bmatrix} R1 \\ R3 \end{bmatrix}$$

This enables one to state the reduced equilibrium equations as follows:

> d8:=evalm(multiply(Kr,Wr)-Rr);

$$d8 := \begin{bmatrix} k11 & W1+k13 & W3-R1 \\ k31 & W1+k33 & W3-R3 \end{bmatrix}$$

Observe that d8 coincides in fact with the first and third equation of d4. The second equation of d4 may then be treated independently.

Consider again the problem of the previous section on the hinged beam made of alternating elements. Noting that $W_1 = W_9 = 0$, first 'unassign' the relevant quantities, load the file autobeam.m and compute the assembly stiffness matrix:

> evaln(Ka):evaln(Kr):evaln(Wr):

126　FINITE ELEMENT METHOD

```
> read 'autobeam.m';
> Kautomat(4,10):
```

MATRIX(((12*EI1/h1**3, 6*EI1/h1**2, −12*EI1/h1**3, 6*EI1/h1**2, 0, 0, 0, 0, 0,0], [6*EI1/h1**2, 4*EI1/h1, −6*EI1/h1**2, 2*EI1/h1, 0, 0, 0, 0, 0, 0], [−12*EI1/h1**3, −6*EI1/h1**2, 12*EI1/h1**3 + 12*EI2/h2**3, −6*EI1/h1**2 + 6*EI2/h2**2, −12*EI2/h2**3, 6*EI2/h2**2, 0, 0, 0, 0], [6*EI1/h1**2, 2*EI1/h1, −6*EI1/h1**2 + 6*EI2/h2**2, 4*EI1/h1 + 4*EI2/h2, −6*EI2/h2**2, 2*EI2/h2, 0, 0, 0, 0], [0, 0, −12*EI2/h2**3, −6*EI2/h2**2, 12*EI2/h2**3 + 12*EI3/h3**3, −6*EI2/h2**2 + 6*EI3/h3**2, −12*EI3/h3**3, 6*EI3/h3**2, 0, 0], [0, 0, 6*EI2/h2**2, 2*EI2/h2, −6*EI2/h2**2 + 6*EI3/h3**2, 4*EI2/h2 + 4*EI3/h3, −6*EI3/h3**2, 2*EI3/h3, 0, 0], [0, 0, 0, 0, −12*EI3/h3**3, −6*EI3/h3**2, 12*EI3/h3**3 + 12*EI4/h4**3, −6*EI3/h3**2 + 6*EI4/h4**2, −12*EI4/h4**3, 6*EI4/h4**2], [0, 0, 0, 0, 6*EI3/h3**2, 2*EI3/h3, −6*EI3/h3**2 + 6*EI4/h4**2, 4*EI3/h3 + 4*EI4/h4, −6*EI4/h4**2, 2*EI4/h4], [0, 0, 0, 0, 0, 0, −12*EI4/h4**3, −6*EI4/h4**2, 12*EI4/h4**3, −6*EI4/h4**2], [0, 0, 0, 0, 0, 0, 6*EI4/h4**2, 2*EI4/h4, −6*EI4/h4**2, 4*EI4/h4]])

Next, compute the reduced matrices and state the reduced equations:

```
> Kr:=submatrix(Ka,[2,3,4,5,6,7,8,10],
  [2,3,4,5,6,7,8,10]):
> Rr:=array([[0],[0],[0],[-P],[0],[0],[0],[0]]):
> Wr:=array([[W2],[W3],[W4],[W5],[W6],[W7],[W8],[W10]]):
> B:=evalm(multiply(Kr,Wr)-Rr):
```

It may be convenient to state the above equations as a set:

```
> s:=array(1..8):
> for i to 8 do
> s[i]:=B[i,1]
> od:
> convert(s,set):
```

This set of equations may be solved in a general form (without displaying the lengthy results):

```
> d14:=solve(",{W2,W3,W4,W5,W6,W7,W8,W10}):
> assign(d14):
```

Substituting for the stiffness of the elements, recover the result of the previous section for W_5 = W5:

```
> subs({EI1=EI,EI2=a*EI,EI3=EI,EI4=a*EI,
> h1=h,h2=h,h3=h,h4=h},W5):
> W5:=simplify(");
```

　　　W5 := −2/3*P*h**3/EI/a*(1 + a)

IMPOSITION OF CONSTRAINTS   127

Observe that this technique has reduced the system of equations from ten to eight.

The next example deals with the clamped composite beam under the force P_0 shown in Fig. F17 (Section F14). Compute first the assembly stiffness matrix, taking into account that the beam has two elements and six degrees of freedom:

> Kautomat(2,6):

MATRIX(((12*EI1/h1**3, 6*EI1/h1**2, − 12*EI1/h1**3, 6*EI1/h1**2, 0, 0],
[6*EI1/h1**2, 4*EI1/h1, − 6*EI1/h1**2, 2*EI1/h1, 0, 0],
[− 12*EI1/h1**3, − 6*EI1/h1**2,12*EI1/h1**3 + 12*EI2/h2**3,
− 6*EI1/h1**2 + 6*EI2/h2**2, − 12*EI2/h2**3, 6*EI2/h2**2],
[6*EI1/h1**2, 2*EI1/h1, − 6*EI1/h1**2 + 6*EI2/h2**2, 4*EI1/h1 + 4*EI2/h2,
− 6*EI2/h2**2, 2*EI2/h2], [0, 0, − 12*EI2/h2**3, − 6*EI2/h2**2,
12*EI2/h2**3, − 6*EI2/h2**2], [0, 0, 6*EI2/h2**2, 2*EI2/h2,
− 6*EI2/h2**2, 4*EI2/h2]])

Since $W_1 = W_2 = W_5 = W_6 = 0$, the reduced matrices are

> Kr:=submatrix(Ka,[3,4],[3,4]);

Kr := MATRIX(((12*EI1/h1**3 + 12*EI2/h2**3, −6*EI1/h1**2
+ 6*EI2/h2**2], [−6*EI1/h1**2 + 6*EI2/h2**2, 4*EI1/h1 + 4*EI2/h2]])

> W3:='W3':W4:='W4':R3:='R3':R4:='R4':
> Wr:=array([[W3],[W4]]);

Wr := MATRIX(((W3], [W4]])

> Rr:=array([[R3],[R4]]);

Rr := MATRIX(((R3], [R4]])

> Pr:=array([[P0/2],[h2*P0/8]]);

Pr := MATRIX(((1/2*P0], [1/8*h2*P0]])

This enables us to state the two reduced equations:

> B:=evalm(multiply(Kr,Wr)-Rr-Pr);

B := MATRIX(((−6*(−2*W3*EI1*h2**3-2*W3*EI2*h1**3 +
W4*h1* h2**3*EI1−W4*h1**3*h2*EI2)/h1**3/h2**3−R3−1/2*P0],
[2*(−3*W3*EI1*h2**2 + 3*W3*EI2*h1**2 + 2*W4*h1*h2**2*EI1
+ 2 *W4*h1**2*h2*EI2)/h1**2/h2**2−R4−1/8*h2*P0]])

> subs({R3=0,R4=0},");

MATRIX(((−6*(−2*W3*EI1*h2**3−2*W3*EI2*h1**3 +
W4*h1*h2**3*EI1−W4*h1**3*h2*EI2)/h1**3/h2**3−1/2*P0],
[2*(−3*W3*EI1*h2**2 + 3*W3*EI2*h1**2 + 2*W4*h1*h2**2*EI1
+ 2*W4*h1**2*h2*EI2)/h1**2/h2**2−1/8*h2*P0]])

128   FINITE ELEMENT METHOD

Solving this system, recover the earlier results:

```
> convert(",set):
> d24:=solve(",{W3,W4}):
> assign(d24):
> W3:=simplify(W3);
```

  W3 := 1/48*(3*EI1*h2**2 + 8*EI1*h1*h2
   + 5*EI2*h1**2)* h2** 3*P0*h1**2/(4*EI2*EI1*h2*h1**3 +
   6*h1**2*EI2*h2**2*EI1 + h1**4*EI2**2 + 4*h1*EI1*h2**3*EI2
   + EI1**2*h2**4)

```
> W4:=simplify(W4);
```

  W4 := 1/8*h1*P0*h2**2*(−EI2*h1**3 + EI1*h2**3
   + 2*h1*EI1* h2**2)/(4*EI2*EI1*h2*h1**3 + 6*h1**2*EI2*h2**2*EI1
   + h1**4*EI2**2 + 4*h1*EI1*h2**3* EI2 + EI1**2*h2**4)

For the homogeneous beam, we trivially get

```
> d26:=subs({EI2=EI1,h2=h1},{W3,W4}):
> simplify(d26[1]);simplify(d26[2]);
```

  1/64*h1**2*P0/EI1
  1/48/EI1*h1**3*P0

The technique of reduced quantities is a convenient method of imposition of constraints, which also extends to the case of nonvanishing nodal values (see Section F26). Sections F36 and F37 present other relevant techniques.

F17. Free vibrations

(MAPLE file vfem39)

In the dynamic case nodal matrices become time-dependent and the equations of motion include the inertia term, as given by (F7.3). The assembly stiffness matrix $[M]_a$ follows from the elemental matrices in the same way as $[K]_a$ does.

Consider free vibrations, in which case the nodal displacement vector $\{W\}_a^{\text{dyn}}$ becomes

$$\{W\}_a^{\text{dyn}} = e^{i\omega t}[W_1\ W_2\ \ldots\ W_N]^T = e^{i\omega t}\{W\}_a \qquad (F17.1)$$

with ω being the natural frequency. Neglecting in (F7.3) the external forces, as the case of free vibrations suggests, and substituting (F17.1) for the displacement, we get

$$(-\omega^2[M]_a + [K]_a)\{W\}_a = 0 \qquad (F17.2)$$

FREE VIBRATIONS 129

Incorporating also the boundary conditions by, say, the method of reduced matrices, one may modify (F17.2) to

$$(-\omega^2[M]_r + [K]_r)\{W\}_r = 0 \qquad (F17.3)$$

which poses a typical eigenvalue problem. Usually, for a system with N degrees of freedom, there are N natural frequencies, which arise as the roots of the determinantal equation. Substituting these roots back in (F17.3) yields a particular mode of vibrations for each of them.

Having in mind free vibrations of a beam and being interested in an automatic derivation, we may adopt the program of Section F14 to compute $[K]_a$ and complement it by a program for computing the assembly mass matrix $[M]_a$.

To this end, below we present a program which automatically computes the assembly mass matrix in a way similar to that of Section F14. Using the definition of the elemental mass matrix given by (F2.4), state $\text{mf} = [m]$,

```
> mf:=array(1..4,1..4):
> for i to 4 do
> for j to 4 do
> if i<=j then
> mf[i,j]:=ro*int(d[i](x)*d[j](x),x=0..h);
> else mf[i,j]:=mf[j,i] fi;
> od;
> od;
```

Again denoting the number of the elements as Z and the number of degrees of freedom as N, introduce the procedure $\text{Mautomat}(Z,N)$, which would compute the assembly mass matrix $\text{Ma} = [M]_a$:

```
> i:='i':j:='j':
> Mautomat:=proc(Z,N)
> for q to Z do mff.q:=array(1..4,1..4):
> for i to 4 do for j to 4 do
> mff.q[i,j]:=subs({ro=ro.q,h=h.q},mf[i,j]):
> od:
> od:
> i:='i':j:='j':
> me.q:=array(1..N,1..N,sparse):
> for i to 4 do for j to 4 do me.q[i+(q-1)*2,
     j+(q-1)*2]:=mff.q[i,j];
> od:
> od:
> od:
> i:='i':j:='j':
> Ma:=array(1..N,1..N):
```

```
> for i to N do for j to N do
> Ma[i,j]:=sum('me.q[i,j]','q'=1..Z):
> od:
> od:
> print(Ma);
> end:
```

For a particular case of the beam consisting of the two elements, the $[K]_a$ has been already derived (Section F16), so we compute herein only $[M]_a$:

```
> Mautomat(2,6):
```

    MATRIX(((13/35*ro1*h1, 11/210*ro1*h1**2, 9/70*ro1*h1,
    − 13/420*ro1*h1**2, 0, 0], [11/210*ro1*h1**2, 1/105*ro1*h1**3,
    13/420*ro1*h1**2, − 1/140*ro1*h1**3, 0, 0], [9/70*ro1*h1,
    13/420*ro1*h1**2, 13/35*ro1*h1 + 13/35*ro2*h2,− 11/210*ro1*h1**2
    + 11/210*ro2*h2**2, 9/70*ro2*h2, − 13/420*ro2*h2**2],
    [− 13/420*ro1*h1**2, − 1/140*ro1*h1**3, − 11/210*ro1*h1**2
    + 11/210*ro2*h2**2, 1/105*ro1*h1**3 + 1/105*ro2*h2**3,
    13/420*ro2*h2**2, − 1/140*ro2*h2**3], [0, 0, 9/70*ro2*h2,
    13/420*ro2*h2**2, 13/35*ro2*h2, − 11/210*ro2*h2**2],
    [0, 0, − 13/420*ro2*h2**2, − 1/140*ro2*h2**3, − 11/210*ro2*h2**2,
    1/105*ro2*h2**3]])

where h1, h2 are the lengths and ro1, ro2 the densities of the elements.

For the clamped beam $W_1 = W_2 = W_5 = W_6 = 0$ and the reduced nodal displacement vector is

$$\{W\}_r = [W_3\ W_4]^T \quad \text{(F17.4)}$$

Consequently, formulate the reduced matrices appearing in (F17.3) as follows:

```
> with(linalg,multiply,genmatrix,submatrix,det):
> Kr:=submatrix(Ka,[3,4],[3,4]);
```

    Kr := MATRIX(((12*EI1/h1**3 + 12*EI2/h2**3, − 6*EI1/h1**2
    + 6*EI2/h2**2], [−6*EI1/h1**2 + 6*EI2/h2**2, 4*EI1/h1 + 4*EI2/h2]])

```
> Mr:=submatrix(Ma,[3,4],[3,4]);
```

    Mr := MATRIX(((13/35*ro1*h1 + 13/35*ro2*h2,
    − 11/210*ro1*h1**2 + 11/210*ro2*h2**2],
    [− 11/210*ro1*h1**2 + 11/210*ro2*h2**2, 1/105*ro1*h1**3
    + 1/105*ro2*h2**3]])

```
> Wr:=array([[W3],[W4]]);
```

    Wr := MATRIX(((W3], [W4]])

FREE VIBRATIONS 131

The system (F17.3) takes the form (wn2 = ω^2):

```
> evalm(-wn2*Mr+Kr):
> d17:=multiply(",Wr);
```

$$d17 := \text{MATRIX}(((1/210*(-78*W3*wn2*h1**4*h2**3*ro1$$
$$-78*W3*wn2*h1**3*h2**4*ro2 + 2520*W3*EI1*h2**3$$
$$+ 2520*W3*EI2*h1**3 + 11*W4*h1**5*h2**3*wn2*ro1$$
$$-11*W4*h1**3*h2**5*wn2*ro2 - 1260*W4*h1*h2**3*EI1$$
$$+ 1260*W4*h1**3*h2*EI2)/h1**3/h2**3],$$
$$[-1/210*(-11*W3*wn2*h1**4*h2**2*ro1 + 11*W3*wn2*h1**2*h2**4*ro2$$
$$+ 1260*W3*EI1*h2**2 - 1260*W3*EI2*h1**2$$
$$+ 2*W4*h1**5*h2**2*wn2*ro1 + 2*W4*h1**2*h2**5*wn2*ro2$$
$$- 840*W4*h1*h2**2*EI1 - 840*W4*h1**2*h2*EI2)/h1**2/h2**2]])$$

For the solution to exist the determinant of this system of homogeneous equations must vanish, which yields the above-mentioned determinantal equation:

```
> convert(",set):
> d18:=genmatrix(",{W3,W4});
```

$$d18 := \text{MATRIX}$$
$$(([-1/210*(2*h1**5*h2**2*wn2*ro1 + 2*h1**2*h2**5*wn2*ro2$$
$$- 840*h1*h2**2*EI1 - 840*h1**2*h2*EI2)/h1**2/h2**2,$$
$$- 1/210*(-11*wn2*h1**4*h2**2*ro1 + 11*wn2*h1**2*h2**4*ro2$$
$$+ 1260*EI1*h2**2 - 1260*EI2*h1**2)/h1**2/h2**2],$$
$$[1/210*(11*h1**5*h2**3*wn2*ro1 - 11*h1**3*h2**5*wn2*ro2$$
$$- 1260*h1*h2**3*EI1 + 1260*h1**3*h2*EI2)/h1**3/h2**3,$$
$$1/210*(-78*wn2*h1**4*h2**3*ro1 - 78*wn2*h1**3*h2**4*ro2$$
$$+ 2520*EI1*h2**3 + 2520*EI2*h1**3)/h1**3/h2**3]])$$

```
> det("):
```

where the display of the expression for the determinant has been prevented.

The `solve` provides the roots, but the expressions are again too complicated to be displayed:

```
> frequency:=solve(",wn2):
```

Compact equations follow for the homogeneous beam, as a particular case,

```
> f:={h2=h1,EI2=EI1,ro2=ro1}:
> subs(f,op(frequency)):map(evalf,");
```

$$\text{op}\left(420.000000\, \frac{EI1}{h1^4\, ro1},\ 32.30769231\, \frac{EI1}{h1^4\, ro1}\right)$$

132 FINITE ELEMENT METHOD

The exact solution is $31.36EI/(\rho h^4)$ for the first natural frequency and $237.9EI/(\rho h^4)$ for the second. Thus, the prediction for the second frequency is rather crude. A finer finite element mesh would provide a better accuracy.

F18. Plate element

We go over to two-dimensional elements. Fig. F19 shows a rectangular plate element. Three components of displacement vector are defined at each of the nodes, namely the deflection w, the rotation θ_x about the x-axis and the rotation θ_y about the y-axis. These quantities are related by

$$\theta_x = -w_{,y}, \qquad \theta_y = w_{,x} \qquad (F18.1)$$

with the positive directions specified by the right-hand screw rule. In this section, we overview basic relations of elementary plate theory. Some of the relevant results are defined more conveniently in the following section with the help of MAPLE.

For the ith node, we specify the displacement vector $\{w\}_i$:

$$\{w\}_i = [w \;\; \theta_x \;\; \theta_y]^T, \qquad i = 1, 2, 3, 4 \qquad (F18.2)$$

which provides a 12×1 nodal displacement matrix for the entire element:

$$\{w\} = [\{w\}_1 \;\{w\}_2 \;\{w\}_3 \;\{w\}_4]^T \qquad (F18.3)$$

As has been noted earlier, the nodal quantities should be defined so as to allow for formulating the strain energy of the structure. Therefore, specify the

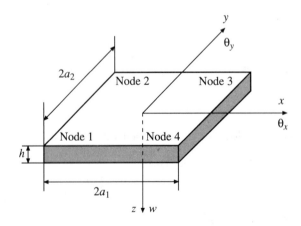

F19 Rectangular plate element

generalized force at the ith node as the matrix $\{q\}_i$

$$\{q\}_i = [q_w \, q_x \, q_y]^T, \qquad i = 1, 2, 3, 4 \tag{F18.4}$$

where q_w is the force and q_x, q_y are the couples. The total nodal force matrix is

$$\{q\} = [\{q\}_1 \, \{q\}_2 \, \{q\}_3 \, \{q\}_4]^T \tag{F18.5}$$

and the relation between $\{q\}$ and $\{w\}$ acquires a familiar form:

$$\{q\} = [k]\{w\} \tag{F18.6}$$

Here $[k]$ is a 12×12 elemental stiffness matrix to be specified.

Now one may proceed with the derivation of the strain energy, which, in turn, leads to the expression for $[k]$. Assume that the shape functions $\{d(x,y)\}$ are known and set, as earlier,

$$w(x, y) = \{d(x,y)\}^T \{w\} \tag{F18.7}$$

The generalized strains $\{\epsilon\}$, according to the elementary plate theory, are

$$\{\epsilon\} = [-w_{,xx} \; -w_{,yy} \; 2w_{,xy}]^T \tag{F18.8}$$

and the generalized stresses $\{\sigma\}$ are

$$\{\sigma\} = [H]\{\epsilon\} \tag{F18.9}$$

Here $[H]$ is the matrix of effective constants:

$$[H] = \frac{Eh^3}{12(1-v^2)} [H]_v \tag{F18.10}$$

where v is Poisson's ratio, h the plate thickness and

$$[H]_v = \begin{bmatrix} 1 & v & 0 \\ v & 1 & 0 \\ 0 & 0 & (1/2 - 1/2v) \end{bmatrix} \tag{F18.11}$$

Substituting (F18.7) in (F18.8) leads to the strain–nodal displacement relation, which we describe symbolically by the matrix $[B]$, referred to henceforth as the *strain–displacement matrix*,

$$\{\epsilon\} = [B]\{w\} \tag{F18.12}$$

The explicit form of $[B]$ will be computed in the following section. Now (F18.9) yields the stress–nodal displacement relation

$$\{\sigma\} = [H][B]\{w\} \tag{F18.13}$$

Given (F18.12) and (F18.13) one may set the relevant energies.

When $\{w\}$ gets a virtual increment $\delta\{w\}$, the deflection gets a virtual increment

$$\delta w = \{d(x,y)\}^T \{\delta w\} \tag{F18.14}$$

as (F18.7) shows. Similarly, (F18.12) yields

$$\delta\{\epsilon\} = [B]\delta\{w\} \quad \text{(F18.15)}$$

and the virtual energy per unit area is

$$\delta\{\epsilon\}^T\{\sigma\} = ([B]\delta\{w\})^T\{\sigma\} = \delta\{w\}^T[B]^T\{\sigma\} \quad \text{(F18.16)}$$

The energy increment is

$$\delta\{w\}^T \iint_S [B]^T\{\sigma\}\,dx\,dy \quad \text{(F18.17)}$$

which should be equal to the work done by the nodal forces

$$\delta\{w\}^T\{q\} \quad \text{(F18.18)}$$

Thus the nodal forces $\{q\}$ are

$$\{q\} = \iint_S [B]^T\{\sigma\}\,dx\,dy \quad \text{(F18.19)}$$

or, taking into account (F18.13),

$$\{q\} = \iint_S [B]^T[H][B]\,dx\,dy\,\{w\} \quad \text{(F18.20)}$$

Comparison of (F18.20) with (F18.6) finally yields the stiffness matrix

$$[k] = \iint_S [B]^T[H][B]\,dx\,dy \quad \text{(F18.21)}$$

As to the distributed force $s(x,y)$, it can be shown in a similar way that (F2.10) extends to

$$\{p\}^T = \iint_S s(x,y)\{d(x,y)\}^T\,dx\,dy \quad \text{(F18.22)}$$

One observes that the above analysis of a two-dimensional element is completely similar to that of the one-dimensional elements. This suggests that three-dimensional elements may also be investigated in the same manner, a feature of a well-defined and consistent method.

F19. Programming for plate element

(MAPLE file vfem40)

In this section we compute explicitly many of the results of the previous section. As earlier, begin with an approximating polynomial for the deflection $w(x,y)$,

PROGRAMMING FOR PLATE ELEMENT

which, according to the number of degrees of freedom, should consist of twelve coefficients,

$$w(x,y) = ao_1 + ao_2 x + ao_3 y + ao_4 x^2 + ao_5 xy + ao_6 y^2 + ao_7 x^3 + ao_8 x^2 y$$
$$+ ao_9 xy^2 + ao_{10} y^3 + ao_{11} x^3 y + ao_{12} xy^3 \qquad (F19.1)$$

where ao_i, $i = 1, 2, \ldots, 12$ are the coefficients. The symbolic computations below do not follow exactly the order of Section F18, as their semantics often allows for a simpler derivation.

State first (F19.1) and its first derivatives wx and wy, using the arrays po and a:

```
> po:=array([[1,x,y,x^2,y^2,x*y,x^3,x^2*y,x*y^2,y^3,
    x^3*y,x*y^3]]):
> a:=array(1..12,1..1):for i to 12 do a[i,1]:=ao[i]: od:
> with(linalg,multiply):
> d5:=multiply(po,a):w:=d5[1,1];
```

$$w := ao[1] + x{*}ao[2] + y{*}ao[3] + x{**}2{*}ao[4] + y{**}2{*}ao[5]$$
$$+ x{*}y{*}ao[6] + x{**}3{*}ao[7] + x{**}2{*}y{*}ao[8] + x{*}y{**}2{*}ao[9]$$
$$+ y{**}3{*}ao[10] + x{**}3{*}y{*}ao[11] + x{*}y{**}3{*}ao[12]$$

```
> wy:=-diff(w,y):wx:=diff(w,x):
```

The above expression for the deflection w may be set as

$$w(x,y) = [po]\{a\} \qquad (F19.2)$$

where $[po]$ and $\{a\}$ are the specified arrays.

Denote the nodal coordinates as xi and yi and the nodal displacement components given by (F18.2) as wi, tetxi and tetyi. Here $i = 1, \ldots, 4$, according to the number of nodes. Below we compute the nodal displacement components and form sequences of them (si) for each node i,

```
> for i to 4 do d6.i:={x=x.i,y=y.i}: w.i:=subs(d6.i,w):
> tetx.i:=subs(d6.i,wy):
> tety.i:=subs(d6.i,wx):
> s.i:=w.i,tetx.i,tety.i:
> od:
```

The above sequences si may be viewed as a system of equations:

$$\{w\} = [c]\{a\} \qquad (F19.3)$$

where $\{a\}$ is the matrix of the coefficients appearing in (F19.2). The genmatrix enables one to easily find $[c] = $ d8

```
> i:='i':
> with(linalg,genmatrix,inverse,transpose):
> co:=ao[i]$ i=1..12;
```

co := ao[1], ao[2], ao[3], ao[4], ao[5], ao[6], ao[7], ao[8], ao[9], ao[10], ao[11], ao[12]

```
> d8:=genmatrix([s1,s2,s3,s4],[co]):
```

Specifying the nodal coordinates in accordance with Fig. F19 (Section F18), we get the final form of $[c]$ = d8:

```
> c:=array(1..12,1..12):
> for i to 12 do for j to 12 do
> c[i,j]:=subs({x1=-a1,y1=-a2,x2=-a1,y2=a2,x3=a1,y3=a2,
  x4=a1,y4=-a2},
> d8[i,j]):
> od:
> od:
```

Note that the shape functions relate $w(x,y)$ to $\{w\}$, as follows:

$$w(x,y) = \{d(x,y)\}^T\{w\} = \{d(x,y)\}^T[c]\{a\} \qquad (F19.4)$$

where use has been made of (F19.3). On the other hand, the deflection $w(x,y)$ has been constructed as a product of po = $[po]$ and a = $\{a\}$, as (F19.2) shows.

Equations (F19.2) and (F19.4) yield a useful expression for the shape functions $\{d(x,y)\}$:

$$\{d(x,y)\}^T = [po][c]^{-1} \qquad (F19.5)$$

The explicit form of $\{d(x,y)\}^T$ = interpoltr is specified below, in agreement with (F19.5):

```
> invc:=inverse(c):interpoltr:=multiply(po,invc):
```

We turn to the generalized strain $\{\epsilon\}$ given by (F18.8) and formulate the matrix $[s]$ = s, which relates $\{\epsilon\}$ to $\{a\}$:

$$\{\epsilon\} = [s]\{a\} \qquad (F19.6)$$

by the following commands:

```
> ex:=-diff(w,x,x):ey:=-diff(w,y,y):exy:=2*diff(w,x,y):
> s:=genmatrix([ex,ey,exy],[co]);
```

s := MATRIX(((0, 0, 0, −2, 0, 0, −6*x, −2*y, 0, 0, −6*x*y, 0], [0, 0, 0, 0, −2, 0, 0, 0, −2*x, −6*y, 0, −6*x*y], [0, 0, 0, 0, 0, 2, 0, 4*x, 4*y, 0, 6*x**2, 6*y**2]])

It follows from (F19.6) and (F19.3) that

$$\{\epsilon\} = [s][c]^{-1}\{w\} \qquad (F19.7)$$

This provides the useful representation for the strain-displacement matrix $[B]$, which relates the strain to the nodal displacement, as given by (F18.12):

$$[B] = [s][c]^{-1} \qquad (F19.8)$$

Compute $[B] = $ B and state the matrix of elastic constants $[H] = $ H, in agreement with (F18.10):

```
> B:=multiply(s,invc):
> H:=evalm(D*array([[1,m,0],[m,1,0],[0,0,m1]]));
```

$$H := \text{MATRIX}(((D, D*m, 0], [D*m, D, 0], [0, 0, D*m1]])$$

where D is the cylindrical stiffness, $m = v$ and $m1 = (1 - v)/2$.

The entries of the elemental stiffness matrix $[k]$ finally follow from (F18.21):

```
> transpose(B):
> g:=multiply(",H,B):
> gc():
> k:=array(symmetric,1..12,1..12): for i to 12 do for j to 12 do
> if i<=j then k[i,j]:=int(int(g[i,j],x=-a1..a1),y=-a2..a2)
> fi;
> od;
> od;
```

For the purpose of illustration, we display below two of the elements of $[k]$:

```
> simplify(k[1,1]);
```

$$1/10*D*(10*a2**4 + 5*m*a1**2*a2**2 + 10*a1**4 + 14*m1*a1**2*a2**2)/a1**3/a2**3$$

```
> simplify(k[1,2]);
```

$$-1/10*D*(2*m1*a2**2 + 10*a1**2 + 5*m*a2**2)/a1/a2**2$$

and then save this matrix and these shape functions as platele.m-file.

```
> save k,interpoltr,'platele.m';
```

F20. Applications

(MAPLE file vfem41)

Consider a square plate, which occupies the domain $-1 \leq x \leq 1$, $-1 \leq y \leq 1$ (Fig. F20). Using a four-element mesh, we first derive the stiffness matrix of the assembly, $[K]_a$. Note that no rotation of axes is needed.

The very process of assembling indicates that the fashion of node numbering influences the structure of equilibrium equations. It is desirable from a computational viewpoint to arrive at a 'banded' matrix, for which non-zero coefficients cluster near its diagonal. This would allow us to take advantage of the sparsity, since zeros outside the band need no processing. By changing the node numbering, one affects the arrangement of nonvanishing entries of the matrix. Generally, it may be stated that a smaller difference among the node

138 FINITE ELEMENT METHOD

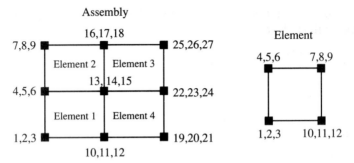

F20 Mesh for a plate

numbers of the element would provide a smaller bandwidth. Account should be taken of this while creating the mesh.

Begin by loading the platele.m-file of Section F19 to get access to the elemental stiffness matrix $[k]$ and shape functions $\{d(x,y)\}$ and then state the assembly displacement vector $\{W\}_a$, which consists of 27 degrees of freedom,

```
> read 'platele.m':
```

and, substituting a1 = a2 = 0.5 and m = v = 0.3, specify $[k]$ for the present case:

```
> K1:=array(symmetric,1..12,1..12):
> for i to 12 do for j to 12 do f[i,j]:=subs({a1=1/2,a2=1/2,
> m=3/10,m1=7/20},k[i,j]):
> K1[i,j]:=simplify(f[i,j]):
> od:
> od:
```

Next, referring to Fig. F20, prepare four expanded matrices Kexi, $i = 1, \ldots, 4$, with the help of connectivities ci,

```
> for i to 4 do Kex.i:=array(1..27,1..27,sparse) od:
> c1:=1,2,3,4,5,6,13,14,15,10,11,12:
> for i to 12 do for j to 12 do Kex1[c1[i],c1[j]]:=K1[i,j] od od:
> c2:=4,5,6,7,8,9,16,17,18,13,14,15:
> for i to 12 do for j to 12 do Kex2[c2[i],c2[j]]:=K1[i,j] od od:
> c3:=13,14,15,16,17,18,25,26,27,22,23,24:
> for i to 12 do for j to 12 do Kex3[c3[i],c3[j]]:=K1[i,j] od od:
> c4:=10,11,12,13,14,15,22,23,24,19,20,21:
> for i to 12 do for j to 12 do Kex4[c4[i],c4[j]]:=K1[i,j] od od:
```

and compute the assembly stiffness matrix $Ka = [K]_a$

> Ka:=evalm(Kex1+Kex2+Kex3+Kex4):

These computations have been performed in a general form, which allows for consideration of a variety of boundary conditions and external loading.

For a clamped plate subjected to a concentrated force Q at its centre, it is convenient to introduce the reduced matrices (see Fig. F20):

> with(linalg,submatrix,transpose,multiply):
> Kr:=submatrix(Ka,[13,14,15],[13,14,15]);

$$Kr := \begin{bmatrix} \dfrac{1056}{25}D & 0 & 0 \\ 0 & \dfrac{152}{25}D & 0 \\ 0 & 0 & \dfrac{152}{25}D \end{bmatrix}$$

> Wr:=submatrix(Wa,[13,14,15],1..1);

$$Wr := \begin{bmatrix} W[13] \\ W[14] \\ W[15] \end{bmatrix}$$

> Rr:=array([[Q],[0],[0]]);

$$Rr := \begin{bmatrix} Q \\ 0 \\ 0 \end{bmatrix}$$

The equilibrium equations and their solution are

> Br:=evalm(multiply(Kr,Wr)-Rr):
> sol:=solve({Br[1,1],Br[2,1],Br[3,1]},{W[13],W[14], W[15]}):
> evalf(sol);

$\{W[14] = 0, W[15] = 0, W[13] = .02367424242\ Q/D\}$

The accurate solution in series yields $W_{13} = 0.0056L^2Q/D$ where L is the plate length. For the present case $L = 2$, which gives $W_{13} = 0.0224Q/D$, reasonably close to the above result.

The next example deals with the uniform load $s(x,y) = \epsilon_0 = e0$. The equivalent nodal forces follow from (F18.22):

$$\{p\} = \iint\limits_{S} s(x,y)\{d(x,y)\}\,dx\,dy \qquad (\text{F20.1})$$

140 FINITE ELEMENT METHOD

We compute these forces below, as an array h[i]

```
> interpol:=transpose(interpoltr):
> for i to 12 do o[i]:=subs({a1=1/2,a2=1/2},interpol[i,1]):
> h[i]:=e0*int(int(o[i],x=-1/2..1/2),y=-1/2..1/2) od:
```

and state the expanded matrices with the help of connectivities:

```
> for i to 4 do Pex.i:=array(1..27,1..1,sparse) od:
> for i to 12 do Pex1[c1[i],1]:=h[i] od:
> for i to 12 do Pex2[c2[i],1]:=h[i] od:
> for i to 12 do Pex3[c3[i],1]:=h[i] od:
> for i to 12 do Pex4[c4[i],1]:=h[i] od:
```

The assembly matrix $Pa = \{P\}_a$ is

```
> Pa:=evalm(Pex1+Pex2+Pex3+Pex4):
```

The reduced displacement vector Wr and reduced stiffness matrix Kr remain the same as in the previous problem:

```
> W[13]:='W[13]':W[14]:='W[14]':W[15]:='W[15]':
> Wr:=array([[W[13]],[W[14]],[W[15]]]):
> Kr:=submatrix(Ka,13..15,13..15):
```

while Pr is

```
> Pr:=submatrix(Pa,13..15,1..1):
```

The equations of equilibrium and their solution are

```
> Br:=evalm(multiply(Kr,Wr)-Pr):
> solve({Br[1,1],Br[2,1],Br[3,1]},{W[13],W[14],W[15]}):
> evalf(");
```

$$\{W[14] = 0, W[15] = 0, W[13] = .02367424242 \frac{e0}{D}\}$$

Within the accuracy of this approximation the result is identical to the previous one.

F21. Two-dimensional elasticity, 1

(MAPLE file vfem42)

Let u and v be the displacement functions along the x and y axes, respectively, and introduce a triangular finite element (Fig. F21). There are two degrees of freedom per node and six of them per element.

Having in mind to derive the stiffness matrix for this element in the context of elasticity theory, begin, as usual, with approximating polynomials, which should have six free coefficients. Therefore u and v are to be linear functions of coordinates, which lead to a constant strain within the triangle.

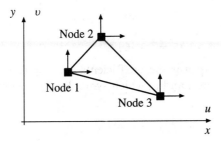

F21 Triangular element

The polynomials for u and v are stated as follows:

```
> Z:=array([[1,x,y]]);
```
$$Z := [1 \; x \; y]$$

```
> Au:=array([[a1],[a2],[a3]]);
```
$$Au := \begin{bmatrix} a1 \\ a2 \\ a3 \end{bmatrix}$$

```
> Av:=array([[a4],[a5],[a6]]);
```
$$Av := \begin{bmatrix} a4 \\ a5 \\ a6 \end{bmatrix}$$

```
> with(linalg,multiply,genmatrix,inverse):
> u:=multiply(Z,Au);
```
$$u := [a1 + x\,a2 + y\,a3]$$

```
> v:=multiply(Z,Av);
```
$$v := [a4 + x\,a5 + y\,a6]$$

As usual, the next step is to represent u and v in terms of the nodal displacement ui and vi, $i = 1, 2, 3$. So, substitute the nodal coordinates xi and yi, $i = 1, 2, 3$ in u and v:

```
> for i to 3 do u.i:=subs({x=x.i,y=y.i},op(u)) od;
```
$$u1 := [a1 + x1\,a2 + y1\,a3]$$
$$u2 := [a1 + x2\,a2 + y2\,a3]$$
$$u3 := [a1 + x3\,a2 + y3\,a3]$$

142   FINITE ELEMENT METHODS

```
> for i to 3 do v.i:=subs({x=x.i,y=y.i},op(v)) od;
```

$$v1 := [a4 + x1\,a5 + y1\,a6]$$
$$v2 := [a4 + x2\,a5 + y2\,a6]$$
$$v3 := [a4 + x3\,a5 + y3\,a6]$$

These equations transform the coefficients `ai`, $i = 1, 2, \ldots, 6$, to the nodal displacement functions u and v, which symbolically may be set as (in MAPLE notation)

$$[u1, u2, u3]^T = cu\,[a1, a2, a3]^T$$
$$[v1, v2, v3]^T = cv\,[a4, a5, a6]^T \tag{F21.1}$$

where the matrices `cu` and `cv` easily follow with the help of `genmatrix`:

```
> cu:=genmatrix([u1[1,1],u2[1,1],u3[1,1]],[a1,a2,a3]);
```

$$cu := \begin{bmatrix} 1 & x1 & y1 \\ 1 & x2 & y2 \\ 1 & x3 & y3 \end{bmatrix}$$

```
> cv:=genmatrix([v1[1,1],v2[1,1],v3[1,1]],[a4,a5,a6]);
```

$$cv := \begin{bmatrix} 1 & x1 & y1 \\ 1 & x2 & y2 \\ 1 & x3 & y3 \end{bmatrix}$$

As anticipated, `cu = cv`. The interpolation functions matrix $\{d(x,y)\}^T =$ `interpol` are just the product of Z-matrix and the inverse of `cu`:

```
> for i to 3 do evaln(u.i): evaln(v.i): od:
> invcu:=inverse(cu):
> s:=multiply(Z,invcu):
> interpol:=map(simplify,op(s));
```

$$\text{interpol} := [\,\frac{x2\,y3 - y2\,x3 - x\,y3 + x\,y2 + y\,x3 - y\,x2}{x2\,y3 - y2\,x3 - x1\,y3 + y1\,x3 + x1\,y2 - y1\,x2},$$
$$-\frac{x1\,y3 - y1\,x3 - x\,y3 + x\,y1 + y\,x3 - y\,x1}{x2\,y3 - y2\,x3 - x1\,y3 + y1\,x3 + x1\,y2 - y1\,x2},$$
$$\frac{x1\,y2 - y1\,x2 - x\,y2 + x\,y1 + y\,x2 - y\,x1}{x2\,y3 - y2\,x3 - x1\,y3 + y1\,x3 + x1\,y2 - y1\,x2}\,]$$

Now the displacement functions u and v take the forms expressed in terms of the nodal quantities:

```
> u:=multiply(s,array([[u1],[u2],[u3]]));
```

$$u := [(u1\,x2\,y3 - u1\,y2\,x3 - u1\,x\,y3 + u1\,x\,y2 + u1\,y\,x3 - u1\,y\,x2$$
$$- u2\,x1\,y3 + u2\,y1\,x3 + u2\,x\,y3 - u2\,x\,y1 - u2\,x3 + u2\,y\,x1$$
$$+ u3\,x1\,y2 - u3\,y1\,x2 - u3\,x\,y2 + u3\,x\,y1 + u3\,y\,x2 - u3\,y\,x1)/$$
$$(x2\,y3 - y2\,x3 - x1\,y3 + y1\,x3 + x1\,y2 - y1\,x2)]$$

```
> v:=multiply(s,array([[v1],[v2],[v3]]));
```

v :=
[-(-v1 x2 y3 + v1 y2 x3 + v1 x y3 - v1 x y2 - v1 y x3 + v1 y x2
+ v2 x1 y3 - v2 y1 x3 - v2 x y3 + v2 x y1 + v2 y x3 - v2 y x1
- v3 x1 y2 + v3 y1 x2 + v3 x y2 - v3 x y1 - v3 y x2 + v3 y x1)/
(x2 y3 - y2 x3 - x1 y3 + y1 x3 + x1 y2 - y1 x2)]

This provides an alternative derivation of the interpolation (shape) functions with the help of `coeff`:

```
> for i to 3 do d[i](x,y):=coeff(collect(u[1,1],u.i),u.i) od;
```

$$d[1](x, y) := \frac{x2\,y3 - y2\,x3 - x\,y3 + x\,y2 + y\,x3 - y\,x2}{x2\,y3 - y2\,x3 - x1\,y3 + y1\,x3 + x1\,y2 - y1\,x2}$$

$$d[2](x, y) := \frac{-x1\,y3 + y1\,x3 + x\,y3 - x\,y1 - y\,x3 + y\,x1}{x2\,y3 - y2\,x3 - x1\,y3 + y1\,x3 + x1\,y2 - y1\,x2}$$

$$d[3](x, y) := \frac{x1\,y2 - y1\,x2 - x\,y2 + x\,y1 + y\,x2 - y\,x1}{x2\,y3 - y2\,x3 - x1\,y3 + y1\,x3 + x1\,y2 - y1\,x2}$$

which is identical to the previous result.

Turning to the stiffness matrix, note that many of the considerations of Section F18 concerning the plate element remain in effect. Particularly, the result for $[k]$ slightly modifies to

$$[k] = \int\!\!\int_S [B]^T [H][B] t \, dx \, dy \qquad (F21.2)$$

where t is the element thickness and $[B]$ is the strain–displacement matrix

$$\{\epsilon\} = [B]\{w\} \qquad (F21.3)$$

with

$$\{w\} = [u1 \ v1 \ u2 \ v2 \ u3 \ v3]^T \qquad (F21.4)$$

To find $[B]$, note that, say, for the state of plane strain,

$$\{\epsilon\} = [u_{,x} \ v_{,y} \ u_{,y} + v_{,x}]^T \qquad (F21.5)$$

The stress matrix and strain matrix are related by

$$\{\sigma\} = [H]\{\epsilon\} \qquad (F21.6)$$

The H-matrix appearing in the above equation is $[H] = E/(1-v^2)H_v$, with H_v being again given by (F18.11), as follows:

$$[H]_v = \begin{bmatrix} 1 & v & 0 \\ v & 1 & 0 \\ 0 & 0 & (1/2 - 1/2v) \end{bmatrix} \qquad (F21.7)$$

144 FINITE ELEMENT METHODS

Going over to computations, first state the components of the strain matrix according to (F21.5). Below $ux = u_{,x}$ etc., also note that in the context of programming u and v are considered as matrices:

```
> ux:=diff(u[1,1],x);vy:=diff(v[1,1],y);
  uyvx:=diff(u[1,1],y)+diff(v[1,1],x);
```

$$ux := \frac{-u1\,y3 + u1\,y2 + u2\,y3 - u2\,y1 - u3\,y2 + u3\,y1}{x2\,y3 - y2\,x3 - x1\,y3 + y1\,x3 + x1\,y2 - y1\,x2}$$

$$vy := -\frac{-v1\,x3 + v1\,x2 + v2\,x3 - v2\,x1 - v3\,x2 + v3\,x1}{x2\,y3 - y2\,x3 - x1\,y3 + y1\,x3 + x1\,y2 - y1\,x2}$$

$$uyvx := \frac{u1\,x3 - u1\,x2 - u2\,x3 + u2\,x1 + u3\,x2 - u3\,x1}{x2\,y3 - y2\,x3 - x1\,y3 + y1\,x3 + x1\,y2 - y1\,x2}$$
$$- \frac{v1\,y3 - v1\,y2 - v2\,y3 + v2\,y1 + v3\,y2 - v3\,y1}{x2\,y3 - y2\,x3 - x1\,y3 + y1\,x3 + x1\,y2 - y1\,x2}$$

In view of (F21.3), genmatrix may provide $[B] = B$. To facilitate the computation, we may refer to the numerators of ux, vy and uyvx only, as their denominator is a constant. More precisely, the denominator is merely the double area of the triangle, which is computed via the above matrix cu

$$S = \det(cu)/2 \tag{F21.8}$$

Indeed,

```
> det(cu)/2;
```

$$1/2\,x2\,y3 - 1/2\,y2\,x3 - 1/2\,x1\,y3 + 1/2\,y1\,x3 + 1/2\,x1\,y2 - 1/2\,y1\,x2$$

Therefore, compute $[B]$ by

```
> w:=u1,v1,u2,v2,u3,v3:
> B:=evalm(genmatrix([numer(ux),numer(vy),numer(uyvx)],
  [w])/(2*S));
```

$$B := [1/2\,\frac{-y3+y2}{S},\ 0,\ 1/2\,\frac{y3-y1}{S},\ 0,\ 1/2\,\frac{-y2+y1}{S},\ 0]$$
$$[0,\ 1/2\,\frac{x3-x2}{S},\ 0,\ 1/2\,\frac{-x3+x1}{S},\ 0\ 1/2\,\frac{x2-x1}{S}]$$
$$[1/2\,\frac{x3-x2}{S},\ 1/2\,\frac{-y3+y2}{S},\ 1/2\,\frac{-x3+x1}{S},\ 1/2\,\frac{y3-y1}{S},$$
$$1/2\,\frac{x2-x1}{S},\ 1/2\,\frac{-y2+y1}{S}]$$

Note, $[B]$ does not depend on the coordinates, so (F21.2) yields for the stiffness matrix

$$[k] = [B]^T[H][B]\,tS \tag{F21.9}$$

TWO-DIMENSIONAL ELASTICITY, 2   145

if the thickness t is a constant. Denoting $H = [H]t$, $et = Et/(1-v^2)$, $m = v$ and $m1 = (1-v)/2$, we get

> H:=evalm(et*array([[1,m,0],[m,1,0],[0,0,m1]]));

$$H := \begin{bmatrix} et & (et\ m) & 0 \\ (et\ m) & et & 0 \\ 0 & 0 & (et\ m1) \end{bmatrix}$$

> kk:=multiply(transpose(B),H):
> k:=multiply(kk,B):

where the display has been prevented. We present below the first entry only

> k[1,1];

$$1/4 \frac{et(y3^2 - 2\,y3\,y2 + y2^2 + m1\,x3^2 - 2\,m1\,x3\,x2 + m1\,x2^2)}{s^2}$$

F22. Two-dimensional elasticity, 2

(MAPLE file vfem42)

Another simple element is a rectangle (Fig. F22).

Its analysis requires only a slight modification compared to the triangular element. The approximating polynomials for the u and v functions should contain four coefficients each, to match eight available degrees of freedom.

> Z:=array([[1,x,y,x*y]]);

   $Z := [1\ x\ y\ x\ y]$

> Au:=array([[a1],[a2],[a3],[a4]]);

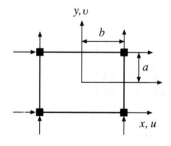

F22 Rectangular element

146 FINITE ELEMENT METHODS

$$Au := \begin{bmatrix} a1 \\ a2 \\ a3 \\ a4 \end{bmatrix}$$

> Av:=array([[a5],[a6],[a7],[a8]]);

$$Au := \begin{bmatrix} a5 \\ a6 \\ a7 \\ a8 \end{bmatrix}$$

> u:=multiply(Z,Au);

    u := [a1 + x a2 + y a3 + x y a4]

> v:=multiply(Z,Av);

    v := [a5 + x a6 + y a7 + x y a8]

Formulate the nodal displacement matrix in terms of the *a*-coefficients:

> for i to 4 do u.i:=subs({x=x.i,y=y.i},op(u)) od;

    u1 := [a1 + x1 a2 + y1 a3 + x1 y1 a4]
    u2 := [a1 + x2 a2 + y2 a3 + x2 y2 a4]
    u3 := [a1 + x3 a2 + y3 a3 + x3 y3 a4]
    u4 := [a1 + x4 a2 + y4 a3 + x4 y4 a4]

> for i to 4 do v.i:=subs({x=x.i,y=y.i},op(v)) od;

    v1 := [a5 + x1 a6 + y1 a7 + x1 y1 a8]
    v2 := [a5 + x2 a6 + y2 a7 + x2 y2 a8]
    v3 := [a5 + x3 a6 + y3 a7 + x3 y3 a8]
    v4 := [a5 + x4 a6 + y4 a7 + x4 y4 a8]

and then derive the interpolation (shape) functions in a way similar to that of the previous section, substituting 2*a* and 2*b* for the size of the elements (Fig. F22). Note that u and v are treated as the matrices

> cu:=genmatrix([u1[1,1],u2[1,1],u3[1,1],u4[1,1]],
  [a1,a2,a3,a4]);

$$cu := \begin{bmatrix} 1 & x1 & y1 & y1 & x1 \\ 1 & x2 & y2 & y2 & x2 \\ 1 & x3 & y3 & x3 & y3 \\ 1 & x4 & y4 & x4 & y4 \end{bmatrix}$$

> for i to 4 do evaln(u.i):evaln(v.i) od:
> invcu:=inverse(cu):
> s:=multiply(Z,invcu):
> u:=multiply(s,array([[u1],[u2],[u3],[u4]])):

```
> v:=multiply(s,array([[v1],[v2],[v3],[v4]])):
> U:=subs({x1=-b,y1=-a,x2=b,y2=-a,x3=b,y3=a,
  x4=-b,y4=a},u[1,1]):
> for i to 4 do f[i](x):=coeff(collect(U,u.i),u.i) od:
> for i to 4 do d[i](x):=simplify(f[i](x)) od;
```

$$d[1](x) := -1/4 \frac{y\,b - x\,y + x\,a - b\,a}{b\,a}$$

$$d[2](x) := 1/4 \frac{x\,a + b\,a - x\,y - y\,b}{b\,a}$$

$$d[3](x) := 1/4 \frac{x\,a + b\,a + x\,y + y\,b}{b\,a} \tag{F22.1}$$

$$d[4](x) := -1/4 \frac{-b\,a + x\,a - y\,b + x\,y}{b\,a}$$

Next, as earlier, state the components of the strain matrix, ux, vy and uyvx

```
> diff(u[1,1],x):ux:=factor(");
```

$$ux := -1/4 \frac{-u1\,y + u4\,y - u2\,a + u1\,a + u4\,a - u3\,y - u3\,a + u2\,y}{b\,a}$$

```
> diff(v[1,1],y):vy:=factor(");
```

$$vy := -1/4 \frac{-v3\,x + v1\,b + v2\,x + v4\,x - v3\,b + v2\,b - v1\,x - v4\,b}{b\,a}$$

```
> diff(u[1,1],y)+diff(v[1,1],x):uyvx:=factor(");
```

$$uyvx := 1/4\,(u1\,x + u4\,b - u4\,x - u1\,b + u3\,x + u3\,b - u2\,x - u2\,b \\ + v3\,y - v2\,y - v1\,a - v4\,a - v4\,y + v3\,a + v2\,a + v1\,y)/(b\,a)$$

which enables one to arrive at the strain–displacement matrix $[B] = $ B, according to (F21.3):

```
> w:=u1,v1,u2,v2,u3,v3,u4,v4:
> B:=genmatrix([ux,vy,uyvx],[w]);
```

$$B := [1/4\frac{-a+y}{a\,b},\ 0,\ 1/4\frac{a-y}{a\,b},\ 0,\ 1/4\frac{y+a}{a\,b},\ 0,\ 1/4\frac{-y-a}{a\,b},\ 0]$$

$$[0,\ 1/4\frac{x-b}{a\,b},\ 0,\ 1/4\frac{-x-b}{a\,b},\ 0,\ 1/4\frac{x+b}{a\,b},\ 0,\ 1/4\frac{b-x}{a\,b}]$$

$$[1/4\frac{x-b}{a\,b},\ 1/4\frac{-a+y}{a\,b},\ 1/4\frac{-x-b}{a\,b},\ 1/4\frac{a-y}{a\,b},\ 1/4\frac{x+b}{a\,b},\ 1/4\frac{y+a}{a\,b},$$

$$1/4\frac{b-x}{a\,b},\ 1/4\frac{-y-a}{a\,b}]$$

148 FINITE ELEMENT METHODS

The integrand of (F21.2) is

```
> kk:=multiply(transpose(B),H):
> integrand:=multiply(kk,B):
```

where H is the same as in the program of Section F21. The stiffness matrix $[k] =$ k is

```
> k:=array(1..8,1..8,symmetric):
> for i to 8 do for j to 8 do
> if i<=j then k[i,j]:=int(int(integrand[i,j],x=-b..b),
  y=-a..a) fi od od;
```

We display below only the first entry

```
> simplify(k[1,1]);
```

$$\frac{1}{3} \frac{et(a^2 + m1\, b^2)}{a\, b}$$

F23. Heat conduction and related problems

On omission of boundary terms (such as convection terms, etc.), the functional

$$F = \iint_S f\, dx\, dy - \int_\delta q_\delta T\, d\delta \qquad (F23.1)$$

where

$$f = k_0/2(T_{,x}^2 + T_{,y}^2) - QT + \rho c_0 T T_{,t} \qquad (F23.2)$$

describes heat conduction in a two-dimensional isotropic medium of unit thickness. The medium occupies a domain S with the boundary δ. Here T, Q, k_0, ρ, c_0, and q_δ are the temperature, rate of heat generation by internal sources, thermal conductivity, mass density, specific heat and prescribed flux, respectively.

To state this time-dependent problem in terms of the finite element method, we refer, for the sake of certainty, to the rectangular four-node element considered in the previous section and set

$$T = [d]\{T\} \qquad (F23.3)$$

where $[d] = [d_1\, d_2\, d_3\, d_4]$ are the shape functions and $\{T\} = [T_1\, T_2\, T_3\, T_4]^T$ the nodal temperature matrix. Indeed, results of Section F22 apply also to any scalar field variable, such as T. Note, the entries of $[d]$ have been given explicitly in Section F22.

The spatial derivatives needed are

$$T_{,x} = [d_{1,x}\ d_{2,x}\ d_{3,x}\ d_{4,x}]\{T\} = [d_{,x}]\{T\}$$
$$T_{,y} = [d_{1,y}\ d_{2,y}\ d_{3,y}\ d_{4,y}]\{T\} = [d_{,y}]\{T\}$$
(F23.4)

and the temporal derivative is

$$T_{,t} = [d]\ [T_{1,t}\ T_{2,t}\ T_{3,t}\ T_{4,t}]^T \tag{F23.5}$$

for the nodal quantities $\{T\}$ are taken as time-dependent.

Since $T^T = T$, we set

$$T_{,x}^2 = T_{,x}^T T_{,x}, \qquad T_{,y}^2 = T_{,y}^T T_{,y} \tag{F23.6}$$

and get for the functional F

$$F = \iint_S [k_0/2(T_{,x}^T T_{,x} + T_{,y}^T T_{,y}) - QT^T + \rho c_0 T^T T_{,t}]\,dx\,dy$$
$$- \int_\delta q_\delta T^T\,d\delta \tag{F23.7}$$

Substituting (F23.3), (F23.4) and (F23.5) in the above equation, we get for this functional

$$F = \iint_S [k_0/2[\{T\}^T[d_{,x}]^T[d_{,x}]\{T\} + \{T\}^T[d_{,y}]^T[d_{,y}]\{T\}] - \{T\}^T[d]^T Q$$
$$+ \rho c_0 \{T\}^T[d]^T[d]\{T_{,t}\}]\,dx\,dy - \int_\delta q_\delta\{T\}^T[d]^T\,d\delta \tag{F23.8}$$

If S is the element area, the above result admits a typical finite element representation:

$$F = 1/2\{T\}^T[k]\{T\} + \{T\}^T[c]\{T_{,t}\} - \{T\}^T\{q\} \tag{F23.9}$$

where

$$[k] = \iint_S k_0([d_{,x}]^T[d_{,x}] + [d_{,y}]^T[d_{,y}])\,dx\,dy$$
$$[c] = \iint_S [d]^T rc_0[d]\,dx\,dy \tag{F23.10}$$
$$\{q\} = \iint_S [d]^T Q\,dx\,dy + \int_\delta [d]^T q_\delta\,d\delta$$

Here $[k]$ may be considered as the 'stiffness' matrix, $\{q\}$ the nodal 'force' matrix and $[c]$ the so-called *capacitance* matrix, which is responsible for transient effects.

150   FINITE ELEMENT METHODS

For the actual process the functional F is stationary, so its variation with respect to $\{T\}$ must vanish:

$$\delta F = F_{,\{T\}} \delta \{T\}^T = 0 \qquad \text{(F23.11)}$$

where $F_{,\{T\}}$ is the derivative of F with respect to the vector $\{T\}$. Making use of the appropriate rule of differentiation of a matrix, reduce the above equation to a familiar form

$$[k]\{T\} + [c]\{T_{,t}\} = \{q\} \qquad \text{(F23.12)}$$

Note that in the case of a steady-state, $T_{,t} = 0$, the Euler–Lagrange equation for the functional F is

$$k_0(T_{,xx} + T_{,yy}) + Q = 0 \qquad \text{(F23.13)}$$

which describes a variety of phenomena, in particular, elastic torsion, if T is a warping function, ground water flow, if T is hydraulic heat, electrostatics, if T is a field potential. Consequently, the finite element formulation applies to these problems too.

F24. The Bubnov–Galerkin formulation

The weighted residual approach, of which the Bubnov–Galerkin method is a particular case, paves the way to the finite element formulation of a wider basis then the direct technique considered so far. Particularly, it fits better problems of heat transfer and fluid flow.

Consider the equation

$$(k_0 T_{,x})_{,x} + (k_0 T_{,y})_{,y} + Q = 0 \qquad \text{(F24.1)}$$

which may describe heat conduction in isotropic two-dimensional medium, among others. The notation used is the same as in Section F23. Equations of this type are closely related to the functional given by (F23.1). The boundary conditions associated with (F24.1) are

$$k_0(n_x T_{,x} + n_y T_{,y}) - q_\delta = 0 \qquad \text{(F24.2)}$$

where n_x and n_y are direction cosines of an outward normal to the boundary surface, δ. In agreement with the basic idea of the Bubnov–Galerkin method, set, instead of (F24.1),

$$\int\int_S [(k_0 T_{,x})_{,x} + (k_0 T_{,y})_{,y} + Q] \delta T \, dx\, dy = 0 \qquad \text{(F24.3)}$$

where δT is the variation of T. Assuming the fundamental finite element representation (F23.3) holds:

$$T = [d]\{T\} \qquad \text{(F24.4)}$$

we get for the variation

$$\delta T = \sum_i d_i(x,y)\delta T_i = \delta\{T\}^T\{d_i(x,y)\} \tag{F24.5}$$

and for (F24.3)

$$\delta\{T\}^T \iint_S \{d_i(x,y)\}[(k_0 T_{,x})_{,x} + (k_0 T_{,y})_{,y} + Q]\,dx\,dy = 0 \tag{F24.6}$$

This suggests that the integral term must vanish and delivers the matrix equation

$$\iint_S \{d_i(x,y)\}[(k_0 T_{,x})_{,x} + (k_0 T_{,y})_{,y} + Q]\,dx\,dy = 0 \tag{F24.7}$$

The further treatment of this expression brings the boundary conditions into play. Recall the formula of integration by parts (Green's theorem):

$$\iint_S F_1 F_{2,x}\,dx\,dy = -\iint_S F_{1,x} F_2\,dx\,dy + \int_\delta F_1 F_2 n_x\,d\delta \tag{F24.8}$$

where F_1 and F_2 are continuous functions of the coordinates. A similar expression holds for the derivative with respect to y. Therefore, (F24.7) becomes

$$\iint_S [-\{d_{,x}\}k_0 T_{,x} - \{d_{,y}\}k_0 T_{,y} + \{d\}Q]\,dx\,dy + \int_\delta \{d\}q_\sigma\,d\delta = 0 \tag{F24.9}$$

where $\{d_{,x}\} = \{d_{1,x} d_{2,x} \cdots\}$ and $\{d_{,y}\} = \{d_{1,y} d_{2,y} \cdots\}$. The boundary term in (F24.9) admits the following representations:

$$\int_\delta \{d\}q_\delta d\delta = \int_\delta \{d\}k_0(n_x T_{,x} + n_y T_{,y})\,d\delta \tag{F24.10}$$

where use has been made of (F24.2). Note that in view of (F24.3)

$$T_{,x} = \{d_{,x}\}^T\{T\},\quad T_{,y} = \{d_{,y}\}^T\{T\} \tag{F24.11}$$

and (F24.9) takes the form

$$\iint_S ([d_{,x}]^T k_0 [d_{,x}] + [d_{,y}]^T k_0 [d_{,y}])\,dx\,dy\{T\}$$
$$= \iint_S \{d\}Q\,dx\,dy + \int_\delta \{d\}q_\delta\,d\delta \tag{F24.12}$$

152 FINITE ELEMENT METHODS

Denoting

$$[k] = \iint_S ([d_{,x}]^T k_0 [d_{,x}] + [d_{,y}]^T k_0 [d_{,y}]) \, dx \, dy$$
$$\{q\} = \iint_S \{d\} Q \, dx \, dy + \int_\delta \{d\} q_\delta \, d\delta \qquad (F24.13)$$

we arrive once again at the elemental equation

$$[k]\{T\} = \{q\} \qquad (F24.14)$$

F25. Heat transfer in a fin

(*MAPLE file vfem43*)

As an application of the above approach, consider a pin fin shown in Fig. F23. The relevant equation is

$$(k_0 S T_{,x})_{,x} - hP(T - T_0) = 0, \qquad 0 \le x \le L \qquad (F25.1)$$

where S, P and L are the cross-sectional area, perimeter and length, respectively. T_0 is the temperature of convection to fluid, h is the heat transfer coefficient. Also we denote the base temperature by T_b. It is seen that (F25.1) is a modification of (F24.1), which accounts for convection. Equation (F25.1) is subject to the boundary conditions

$$T(x = 0) = T_b, \qquad T_{,x}(x = L) = 0 \qquad (F25.2)$$

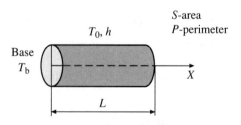

F23 To heat transfer in a pin fin

HEAT TRANSFER IN A FIN

Considering the problem as one-dimensional and denoting the two adjacent nodes as x_1 and x_2, write the basic equation of the Bubnov–Galerkin method, which is a version of (F24.7),

$$\int_{x_1}^{x_2} \{d(x)\}[(k_0 S T_{,x})_{,x} - hP(T - T_0)]\,dx = 0 \tag{F25.3}$$

where the one-dimensional shape functions $\{d(x)\}$ are

$$d_1(x) = (x_2 - x)/(x_2 - x_1), \qquad d_2(x) = (x - x_1)/(x_2 - x_1) \tag{F25.4}$$

Integrating by parts the first term in (F25.3), in agreement with the procedure of Section F24, we get

$$\int_{x_1}^{x_2} \{d(x)\}(k_0 S T_{,x})_{,x}\,dx = \{d(x)\}k_0 S T_{,x}\big|_{x_1}^{x_2} - \int_{x_1}^{x_2} \{d_{,x}(x)\}k_0 S T_{,x}\,dx \tag{F25.5}$$

For the interior points, the integrated term on the right-hand side cancels out during the assembling, so only the last term should be preserved for further computations. Equation (F25.3) takes the form

$$\int_{x_1}^{x_2} \{d_{,x}(x)\}k_0 S T_{,x}\,dx + \int_{x_1}^{x_2} \{d(x)\}hP(T - T_0)\,dx = 0 \tag{F25.6}$$

Using once again the representation (F24.4)

$$T = [d]\{T\} \tag{F25.7}$$

we get (F24.14), namely $[k]\{T\} = \{q\}$, where

$$[k] = [k]^x + [k]^c$$

$$[k]^x = \int_{x_1}^{x_2} \{d_{,x}(x)\}k_0 S \{d_{,x}(x)\}^T\,dx$$

$$[k]^c = \int_{x_1}^{x_2} \{d(x)\}hP\{d(x)\}^T\,dx \tag{F25.8}$$

$$\{q\} = \int_{x_1}^{x_2} \{d(x)\}hPT_0\,dx$$

Note that k_0, h, S, T_0 and P may, in general, depend on x, which would complicate the integration. In view of the concept of 'smallness' inherent in the definition of the finite element, we treat these quantities as constants, which simplifies the analysis.

Turning to symbolic computations, specify the element nodes as $x_1 = 0$ and $x_2 = L_e$ and state the shape functions (F25.4) and their derivatives, $d = \{d\}$ and $dx = \{d_{,x}\}$,

```
> f1:=(Le-x)/Le:f2:=x/Le:
> d:=array([[f1],[f2]]);
```

$$d := \begin{bmatrix} \dfrac{Le - x}{Le} \\ \dfrac{x}{Le} \end{bmatrix}$$

```
> with(linalg,transpose,multiply,submatrix):
> dx:=map(diff,op(d)x);
```

$$dx := \begin{bmatrix} -\dfrac{1}{Le} \\ \dfrac{1}{Le} \end{bmatrix}$$

Next, compute the integrals appearing in (F25.8) $\mathrm{kx} = [k]^x$ (without the constant multiplier) and $\mathrm{kc} = [k]^c$ (without the constant multiplier):

```
> kx:=map(int,multiply(dx,transpose(dx)),x=0..Le);
```

$$kx := \begin{bmatrix} \dfrac{1}{Le} & -\dfrac{1}{Le} \\ -\dfrac{1}{Le} & \dfrac{1}{Le} \end{bmatrix}$$

```
> kc:=map(int,multiply(d,transpose(d)),x=0..Le);
```

$$kc := \begin{bmatrix} 1/3Le & 1/6Le \\ 1/6Le & 1/3Le \end{bmatrix}$$

```
> q:=map(int,op(d),x=0..Le);
```

$$q := \begin{bmatrix} 1/2\,Le \\ 1/2\,Le \end{bmatrix}$$

Note that this integration is trivial. The above coefficients will be recovered in the next section while dealing with a numerical example.

F26. Numerical example

(MAPLE file vfem43)

Fig. F24 shows a modelling of a circular fin by the two finite elements. The values of the relevant parameters are taken as $k_0 = 400\,\mathrm{W/m\,°C}$, $h = 150\,\mathrm{W/m^2\,°C}$, $T_0 = 20°\mathrm{C}$, $L = 0.02\,\mathrm{m}$ and $R = D/2 = 0.2\,\mathrm{cm}$. The base temperature is $T_b = 80°\mathrm{C}$, which is also the temperature at node 1.

NUMERICAL EXAMPLE 155

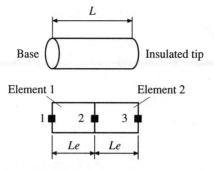

F24 Modeling a pin fin

Compute first the numerical values of the elemental quantities kx, kc and q, as given in Section F25, for the elemental length Le = 0.01 m, and state the above values of the parameters:

> kx:=subs(Le=0.01,op(kx));

$$kx := \begin{bmatrix} 100. & -100. \\ -100. & 100. \end{bmatrix}$$

> kc:=subs(Le=0.01,op(kc));

$$kc := \begin{bmatrix} .003333333333 & .001666666667 \\ .001666666667 & .003333333333 \end{bmatrix}$$

> q:=subs(Le=0.01,op(q));

$$q := \begin{bmatrix} .005000000000 \\ .005000000000 \end{bmatrix}$$

> Le:=0.01:R:=0.002:S:=evalf(Pi*R^2);P:=evalf(Pi*2*R);
  k0:=400:T0:=20:

$$S := .00001256637062$$

> h:=150:

Next, recover the multipliers, omitted in the last computations of Section F25, with the notation k1x = $[k]^x$, k1c = $[k]^c$ and using (F25.8):

> k1x:=evalm(k*S*kx);

$$k1x := \begin{bmatrix} .5026548248 & -.5026548248 \\ -.5026548248 & .5026548248 \end{bmatrix}$$

> k1c:=evalm(h*P*kc);

$$k1c := \begin{bmatrix} .006283185309 & .003141592656 \\ .003141592656 & .006283185309 \end{bmatrix}$$

156 FINITE ELEMENT METHODS

This enables one to formulate the local matrices for the elements 1 and 2:

```
> k1:=evalm(k1x+k1c);
```

$$k1 := \begin{bmatrix} .5089380101 & -.4995132321 \\ -.4995132321 & .5089380101 \end{bmatrix}$$

```
> q1:=evalm(h*P*T0*q);
```

$$q1 := \begin{bmatrix} .1884955593 \\ .1884955593 \end{bmatrix}$$

and the expanded matrices with the help of connectivity:

```
> K1ex:=array(1..3,1..3,sparse):K2ex:=array(1..3,1..3,
  sparse):
> R1ex:=array(1..3,1..1,sparse):R2ex:=array(1..3,1..1,
  sparse):
> for i to 2 do for j to 2 do K1ex[i,j]:=k1[i,j] od
  od:print(K1ex);
```

$$\begin{bmatrix} .508938010 & -.4995132321 & 0 \\ -.4995132321 & .5089380101 & 0 \\ 0 & 0 & 0 \end{bmatrix}$$

```
> c2:=2,3:
> for i to 2 do for j to 2 do K2ex[c2[i],c2[j]]:=k1[i,j] od od:
> print(K2ex);
```

$$\begin{bmatrix} 0 & 0 & 0 \\ 0 & .5089380101 & -.4995132321 \\ 0 & -.4995132321 & .5089380101 \end{bmatrix}$$

```
> for i to 2 do R1ex[i,1]:=q1[i,1] od:print(R1ex);
```

$$\begin{bmatrix} .1884955593 \\ .1884955593 \\ 0 \end{bmatrix}$$

```
> for i to 2 do R2ex[c2[i],1]:=q1[i,1] od:print(R2ex);
```

$$\begin{bmatrix} 0 \\ .1884955593 \\ .1884955593 \end{bmatrix}$$

The assembly matrices are

```
> Ka:=evalm(K1ex+K2ex);
```

$$Ka := \begin{bmatrix} .5089380101 & -.4995132321 & 0 \\ -.4995132321 & 1.017876020 & -.4995132321 \\ 0 & -.4995132321 & .5089380101 \end{bmatrix}$$

NUMERICAL EXAMPLE   157

```
> Ra:=evalm(R1ex+R2ex);
```

$$Ra := \begin{bmatrix} .1884955593 \\ .3769911186 \\ .1884955593 \end{bmatrix}$$

```
> Ta:=array([[T1],[T2],[T3]]);
```

$$Ta := \begin{bmatrix} T1 \\ T2 \\ T3 \end{bmatrix}$$

This enables one to set the assembly equations

$$[K]_a \{T\}_a = \{R\}_a \tag{F26.1}$$

Note that one of the nodal temperatures is given, $T_1 = 80°C$. The solution could be obtained by substituting this in (F26.1) and resolving the system for T_2 and T_3. A more explicit way is to merely replace the first equation in (F26.1) by $T_1 = 80$, redefining properly $[K]_a = $ Ka, and $\{R\}_a = $ Ra,

```
> Ka[1,1]:=1:Ka[1,2]:=0:Ka[1,3]:=0:print(Ka);
```

$$\begin{bmatrix} 1 & 0 & 0 \\ -.4995132321 & 1.017876020 & -.4995132321 \\ 0 & -.4995132321 & .5089380101 \end{bmatrix}$$

```
> Ra[1,1]:=80:print(Ra);
```

$$\begin{bmatrix} 80 \\ .3769911186 \\ .1884955593 \end{bmatrix}$$

The assembly equations and their solution are

```
> Q:=evalm(multiply(Ka,Ta)-Ra);
```

$$Q := \begin{bmatrix} T1 - 80 \\ -.4995132321\,T1 + 1.017876020\,T2 - .4995132321\,T3 - .3769911186 \\ -.4995132321\,T2 + .5089380101\,T3 - .1884955593 \end{bmatrix}$$

```
> s:=Q[y,1] $y=1..3:
> solve({s},{T1,T2,T3});
```

$$\{T2 = 76.80449885, T1 = 80., T3 = 75.75256369\}$$

The technique of reduced quantities introduced in Section F16 may be so modified as to include the case of nonvanishing nodal values. To this end, write explicitly the relevant system

$$k_{11}T_1 + k_{12}T_2 + k_{13}T_3 = R_1$$
$$k_{21}T_1 + k_{22}T_2 + k_{23}T_3 = R_2 \quad \text{(F26.2)}$$
$$k_{31}T_1 + k_{32}T_2 + k_{33}T_3 = R_3$$

and assume that, say,

$$T_1 = T^* \quad \text{(F26.3)}$$

Then the second and third equations of the system may be set as

$$k_{22}T_2 + k_{23}T_3 = R_2 - k_{21}T^*$$
$$k_{32}T_2 + k_{33}T_3 = R_3 - k_{31}T^* \quad \text{(F26.4)}$$

Solving (F26.4), one gets T_2 and T_3. Here are the computations:

> Kr:=submatrix(Ka,[2,3],[2,3]);

$$\text{Kr} := \begin{bmatrix} 1.017876020 & -.4995132321 \\ -.4995132321 & .5089380101 \end{bmatrix}$$

> Tr:=submatrix(Ta,[2,3],1..1);

$$\text{Tr} := \begin{bmatrix} T2 \\ T3 \end{bmatrix}$$

> Rr:=array([[Ra[2,1]-Ka[2,1]*80],[Ra[3,1]-Ka[3,1]*80]]);

$$\text{Rr} := \begin{bmatrix} 40.33804969 \\ .1884955593 \end{bmatrix}$$

> Qr:=evalm(multiply(Kr,Tr)-Rr);

$$\text{Qr} := \begin{bmatrix} -40.33804969 + 1.017876020\,T2 - .4995132321\,T3 \\ -.4995132321\,T2 + .5089380101\,T3 - .1884955593 \end{bmatrix}$$

> solve({Qr[1,1],Qr[2,1]},{T2,T3});

$$\{T2 = 76.80449885, T3 = 75.75256369\}$$

which reproduces the previous results.

F27. Time-history analysis

(*MAPLE file vfem43*)

In case of the transient state, the dimensionality increases by one, since time becomes an additional 'coordinate'. To take account of time-dependence, one may set, in the case, say, of heat transfer,

$$T(x,y,z,t) = [d(x,y,z,t)]\{T\} \quad \text{(F27.1)}$$

which assumes the time-dependent shape functions or, alternatively,

$$T(x,y,z,t) = [d(x,y,z)]T(t) \tag{F27.2}$$

which suggests the time-dependent nodal quantity (temperature). Note that in the analysis of free vibrations in Section F17 and that of heat transfer in Section F23 the nodal quantities have been assumed time-dependent, in agreement with (F27.2). We adopt the same approach herein, considering finite difference schemes for solution of transients. The subject, in general, is quite involved and require an extensive treatment. We present below only a basic approach (see also Chapter S).

The temporal derivative of $\{T(t)\}$ admits various finite difference representations. For example, the forward difference

$$\{T_{,t}\} = (\{T(t+\Delta t)\} - \{T(t)\})/\Delta t \tag{F27.3}$$

or the backward difference

$$\{T_{,t}\} = (\{T(t)\} - \{T(t-\Delta t)\})/\Delta t \tag{F27.4}$$

Here Δt is the time step, which is small but nonzero.

Generalizing the elemental equation (F23.12) to the assembly case, we get

$$[K]_a\{T\}_a + [C]_a\{T_{,t}\}_a = \{R(t)\}_a \tag{F27.5}$$

Together with (F27.3) this yields (for simplicity we omit the label 'a')

$$[C](\{T(t+\Delta t)\} - \{T(t)\})/\Delta t + [K]\{T(t)\} = \{R(t)\} \tag{F27.6}$$

and together with (F27.4) this yields

$$[C](\{T(t)\} - \{T(t-\Delta t)\})/\Delta t + [K]\{T(t)\} = \{R(t)\} \tag{F27.7}$$

Introducing the subscripts i and $i+1$ for the subsequent instants of time, we get more convenient forms:

$$[C]\{T\}_{i+1} = [C - K\Delta t]\{T\}_i + \{R\}_i \Delta t \tag{F27.8}$$

for (F27.6) and

$$[C + K\Delta t]\{T\}_{i+1} = [C]\{T\}_i + \{R\}_{i+1}\Delta t \tag{F27.9}$$

for (F27.7). These equations show that the nodal temperature values may be evaluated recursively, if the initial conditions are given.

A useful generalization of the above equations is the following expression:

$$[C + \alpha K\Delta t]\{T\}_{i+1} = [C - (1-\alpha)K\Delta t]\{T\}_i + [(1-\alpha)\{R\}_i + \alpha\{R\}_{i+1}]\Delta t \tag{F27.10}$$

which yields, as particular cases, the forward or backward difference schemes, if α takes on the value 0 or 1, respectively. Equation (F27.10) also provides the so-called central difference scheme, if $\alpha = 1/2$.

For the time-independent $\{R\}$, the equation (F27.10) may be rewritten as

$$[K]^{\text{eff}}\{T\}_{i+1} = \{R\}_i^{\text{eff}} \tag{F27.11}$$

with

$$\begin{aligned} [K]^{\text{eff}} &= [C + \alpha K \Delta t] \\ \{R\}_i^{\text{eff}} &= [C - (1-\alpha) K \Delta t]\{T\}_i + \{R\}\Delta t \end{aligned} \tag{F27.12}$$

For the given time step Δt, (F27.11) states a recurrence scheme, which may yield the transient response. The smaller Δt, the more accurate results are obtained.

The relevant matrices $[K]$ and $\{R\}$ for the circular fin have been found in Section F26. It remains to specify the capacitance matrix $[C]$. Modifying (F23.10), we get

$$[c] = \int_0^{L_e} [d]^T \rho c_0 S [d] \, dx \tag{F27.13}$$

Compute first the analytical expression for $[c] = c$, using the notation $\rho = \text{ro}$ and $c_0 = \text{c0}$,

```
> Le:='Le':S:='S':
> map(int,multiply(d,transpose(d)),x=0..Le):
> c:=evalm("*ro*c0*S);
```

$$c := \begin{bmatrix} 1/3 \text{ ro c0 S Le} & 1/6 \text{ ro c0 S Le} \\ 1/6 \text{ ro c0 S Le} & 1/3 \text{ ro c0 S Le} \end{bmatrix}$$

Going over to a numerical example, note the crucial role played by the value of the time step Δt, which may affect both the accuracy and stability of the computations.

F28. Numerical example

(MAPLE file vfem43)

We consider a transient state of the pin fin investigated in Section F26. It is suggested that its initial temperature is 20°C. Then the base (node 1) gets the temperature $T_1 = 80°C$ and after the transient process the fin reaches the steady-state case dealt with in Section F26, where the numerical value of governing parameters can be found. In addition, we need the values of the density and specific heat. These are $\rho = 8900 \text{ kg/m}^3$ and $c_0 = 375 \text{ J/kg°C}$, respectively.

NUMERICAL EXAMPLE

For convenience, we reproduce herein the relevant equations of the previous section, (F27.11) and (F27.12)

$$[K]^{\text{eff}}\{T\}_{i+1} = \{R\}_i^{\text{eff}} \tag{F28.1}$$

with

$$[K]^{\text{eff}} = [C + \alpha K \Delta t]$$
$$\{R\}_i^{\text{eff}} = [C - (1-\alpha)K\Delta t]\{T\}_i + \{R\}\Delta t \tag{F28.2}$$

and the subscript 'a' omitted.

Continuing the programming, compute first the values of the capacitance matrices:

```
> c:=subs({Le=0.01,c0=375,ro=8900,S=evalf(Pi*R^2)},
  op(c));
```

$$c := \begin{bmatrix} .1398008731 & .06990043658 \\ .06990043658 & .1398008731 \end{bmatrix}$$

The expanded capacitance matrices are obtained with the help of connectivity:

```
> C1ex:=array(1..3,1..3,sparse):
  C2ex:=array(1..3,1..3,sparse):
> for i to 2 do for j to 2 do C1ex[i,j]:=c[i,j] od od:
> c2:=2,3:
> for i to 2 do for j to 2 do C2ex[c2[i],c2[j]]:=c[i,j] od od:
```

which yields the assembly capacitance matrix $[C]_a = $ Ca:

```
> Ca:=evalm(C1ex+C2ex);
```

$$Ca := \begin{bmatrix} .1398008731 & .06990043658 & 0 \\ .06990043658 & .2796017462 & .06990043658 \\ 0 & .06990043658 & .1398008731 \end{bmatrix}$$

Specify the time step $\text{step} = \Delta t = 0.1$ s and $\text{alpha} = \alpha = \frac{2}{3}$ and recompute the assembly matrices Ka and Ra:

```
> step:=0.1:alpha:=2/3:
> Ka:=evalm(K1ex+K2ex):Ra:=evalm(R1ex+R2ex):
```

which yields $[K]^{\text{eff}} = $ Keff according to (F28.2):

```
> Keff:=evalm(Ca+alpha*step*Ka);
```

$$\text{Keff} := \begin{bmatrix} .1737300738 & .03659955444 & 0 \\ .03659955444 & .3474601475 & .03659955444 \\ 0 & .03659955444 & .1737300738 \end{bmatrix}$$

Referring to (F28.1), state the initial nodal temperatures $\text{Ta} = \{T\}_a$ and compute the initial $\text{Reff0} = R_a^{\text{eff}}$:

162   FINITE ELEMENT METHODS

```
>with(linalg,linsolve,vector):
>Ta0:=array([[20],[20],[20]]):
>sRa:=evalm(step*Ra);
```

$$sRa := \begin{bmatrix} .01884955593 \\ .03769911186 \\ .01884955593 \end{bmatrix}$$

```
>r:=evalm(Ca-(1-alpha)*step*Ka):
>Reff0:=evalm(multiply(r,Ta0)+sRa);
```

$$Reff0 := \begin{bmatrix} 4.206592565 \\ 8.413185128 \\ 4.206592565 \end{bmatrix}$$

For the next instant of time $t = 0.1$ s enforce the condition $T_1 = 80°C$. To this end use the technique intoduced in Sections F16 and F26 (see in particular the discussion on (F26.3) and (F26.4)). The matrices Reff0 and Keff are modified as shown below:

```
>Reff0:=array([[80],[Reff0[2,1]
 -Keff[2,1]*80],[Reff0[3,1]-Keff[3,1]*80]]);
```

$$Reff0 := \begin{bmatrix} 80 \\ 5.485220773 \\ 4.206592565 \end{bmatrix}$$

```
>Keff[1,1]:=1:Keff[1,2]:=0:Keff[1,3]:=0:Keff[2,1]:=0:
 Keff[3,1]:=0:
>print(Keff);
```

$$\begin{bmatrix} 1 & 0 & 0 \\ 0 & .3474601475 & .03659955444 \\ 0 & .03659955444 & .1737300738 \end{bmatrix}$$

The solution of (F28.1), which would provide Ta1 $= \{T(t = 0.1)\}_a$, follows by linsolve. This command may work better when its second argument is defined as a vector. We therefore convert this nodal quantity to a vector when needed:

```
>n:=convert(Reff0,vector):
>linsolve(Keff,n):Ta.1:=convert(",matrix);
```

$$Ta1 := \begin{bmatrix} 80 \\ 13.53649783 \\ 21.36166005 \end{bmatrix}$$

Note that the result for T_2 seems physically meaningless and may follow from the approximation involved. Using a smaller step time and/or different value of α, one may improve this prediction.

Going over to the next instants of time, use the do-command to perform the relations formulated in (F28.1) and (F28.2). Recompute first $[K]^{\text{eff}}$, denoting this value as Kef:

```
> Kef:=evalm(Ca+alpha*step*Ka);
```

$$\text{Kef} := \begin{bmatrix} .1737300738 & .03659955444 & 0 \\ .03659955444 & .3474601475 & .03659955444 \\ 0 & .03659955444 & .1737300738 \end{bmatrix}$$

Then, enforcing the condition $T_1 = 80$ by the above technique, calculate iteratively $\text{Tai} = \{T(t=0.1)\}_a$, up to say $t = 7$ s:

```
> for i to 69 do Reff.i:=evalm(multiply(r,Ta.i)+sRa):
> v.i:=vector(3,[80,Reff.i[2,1]-Kef[2,1]*80,
  Reff.i[3,1]-Kef[3,1]*80]):
> V.(i+1):=linsolve(Keff,v.i):Ta.(i+1):=convert(V.(i+1),
  matrix): od:
```

To observe the evolution of nodal temperatures in time, we present below the results for $t = 1$ s, $t = 3$ s and $t = 7$ s for T_1, T_2 and T_3:

```
> [Ta10[1,1],Ta30[1,1],Ta70[1,1]];
```

[80., 80., 80.]

```
> [Ta10[2,1],Ta30[2,1],Ta70[2,1]];
```

[52.317, 71.997, 76.631]

```
> [Ta10[3,1],Ta30[3,1],Ta70[3,1]];
```

[41.133, 68.949, 75.500]

It is seen that the nodal temperatures indeed approach those of the steady-state investigated in Section F26.

F29. Natural reference systems

(MAPLE file vfem44)

The use of specific coordinate systems, which are adjusted to a particular finite element at hand, may significantly simplify the analysis and bring about more general formulations. In a natural coordinate system any point of the element is specified by dimensionless numbers, whose magnitude does not exceed unity.

A one-dimensional lineal element is shown in Fig. F25(a). Denote the natural (length) coordinates of a point A as (L_1, L_2) and its cartesian coordinate as X:

```
> cartesian:=array([[1],[X]]);
```

164 FINITE ELEMENT METHODS

(a)
```
     X1              X2
     ■───────────────▶
     1,0             0,1
```

(b)
```
     X1      X3      X2
     ■───────■───────▶
     -1      0       1
```

F25 Lineal elements

$$\text{cartesian} := \begin{bmatrix} 1 \\ X \end{bmatrix}$$

> natural:=array([[L1],[L2]]);

$$\text{natural} := \begin{bmatrix} L1 \\ L2 \end{bmatrix}$$

and the operator of mapping (L_1, L_2) to X as transform:

> transform:=array([[1,1],[X1,X2]]);

$$\text{transform} := \begin{bmatrix} 1 & 1 \\ X1 & X2 \end{bmatrix}$$

Then the function $X = X(L_1, L_2)$ is given by

> with(linalg,multiply,inverse,transpose):
> op(cartesian)=multiply(transform,natural);

$$\begin{bmatrix} 1 \\ X \end{bmatrix} = \begin{bmatrix} L1 + L2 \\ X1\,L1 + X2\,L2 \end{bmatrix}$$

The first of the above equations merely states that $L_1 + L_2 = 1$, while the second yields the transformation law

$$X = L_1 X_1 + L_2 X_2 \tag{F29.1}$$

To find the inverse relations $L_1(X)$ and $L_2(X)$, enter

> d7:=inverse(transform);

$$d7 := \begin{bmatrix} \dfrac{X2}{X2 - X1} & -\dfrac{1}{X2 - X1} \\ -\dfrac{X1}{X2 - X1} & \dfrac{1}{X2 - X1} \end{bmatrix}$$

> op(natural)=multiply(d7,cartesian);

$$\begin{bmatrix} L1 \\ L2 \end{bmatrix} = \begin{bmatrix} -\dfrac{-X2 + X}{X2 - X1} \\ \dfrac{-X1 + X}{X2 - X1} \end{bmatrix}$$

Consequently,

$$L_1 = (X_2 - X)/(X_2 - X_1), \qquad L_2 = (X - X_1)/(X_2 - X_1) \tag{F29.2}$$

The endpoints X_1 and X_2 therefore have the natural coordinates $L_1 = 1$, $L_2 = 0$ and $L_1 = 0$, $L_2 = 1$, respectively.

Instead of the two interdependent coordinates, L_1 and L_2, one may use the single natural coordinate L (see Fig. F25(b)), which relates to X either by

```
> X=(1-L)*X1/2+(L+1)*X2/2;
```

$$X = 1/2\,(1 - L)\,X1 + 1/2\,(L + 1)\,X2 \tag{F29.3}$$

or by the inverse

```
> solve(",L):
> L:=simplify(");
```

$$L := \frac{2X - X1 - X2}{X2 - X1} \tag{F29.4}$$

Turning to a triangular element, as shown in Fig. F26, define the area coordinates by

$$L_1 = A_1/A, \qquad L_2 = A_2/A, \qquad L_3 = A_3/A \tag{F29.5}$$

where A is the element area and A_i, $i = 1, 2, 3$ are the areas of the smaller triangles shown: A_1 = area 23c, A_2 = area 3c1, A_3 = area 12c. Note that c is an arbitrary point, not a node. Only two of the area coordinates are independent.

Key in the cartesian and area coordinates and then the relevant transformation

```
> cartesian:=array([[1],[X],[Y]]);
```

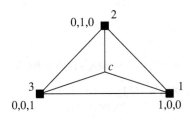

F26 Triangular element

$$\text{cartesian} := \begin{bmatrix} 1 \\ X \\ Y \end{bmatrix}$$

```
> natural:=array([[L1],[L2],[L3]]);
```

$$\text{natural} := \begin{bmatrix} L1 \\ L2 \\ L3 \end{bmatrix}$$

```
> transform:=array([[1,1,1],[X1,X2,X3],[Y1,Y2,Y3]]);
```

$$\text{transform} := \begin{bmatrix} 1 & 1 & 1 \\ X1 & X2 & X3 \\ Y1 & Y2 & Y3 \end{bmatrix}$$

where X_i and Y_i, $i = 1, 2, 3$ are the nodal coordinates. Then the cartesian and area coordinates are related by

```
> op(cartesian)=multiply(transform,natural);
```

MATRIX([[1], [X], [Y]]) = MATRIX([[L1 + L2 + L3], [X1*L1 + X2*L2 + X3*L3], [Y1*L1 + Y2*L2 + Y3*L3]])

These relations show that

$$L_1 + L_2 + L_3 = 1, \quad X = L_1 X_1 + L_2 X_2 + L_3 X_3, \quad Y = L_1 Y_1 + L_2 Y_2 + L_3 Y_3$$
(F29.6)

The inverse is as follows:

```
> d18:=inverse(transform):
> d19:=op(natural)=multiply(d18,cartesian);
```

d19 := MATRIX([[L1], [L2], [L3]]) = MATRIX([[− (X2*Y3 − X3*Y2 − X*Y3 + X*Y2 + Y*X3 − Y*X2)/
(− X2*Y3 + X3*Y2 + X1*Y3 − X1*Y2 − Y1*X3 + Y1*X2)],
[(X1*Y3 − Y1*X3 − X*Y3 + X*Y1 + Y*X3 − Y*X1)/
(−X2*Y3 + X3*Y2 + X1*Y3 − X1*Y2 − Y1*X3 + Y1*X2)],
[− (X1*Y2 − Y1*X2 − X*Y2 + X*Y1 + Y*X2 − Y*X1)/
(− X2*Y3 + X3*Y2 + X1*Y3 − X1*Y2 − Y1*X3 + Y1*X2)]])

In particular, the area coordinates of the nodes are

```
> d20:=subs({X=X1,Y=Y1},op(2,d19)):
  op(natural)=map(simplify,op(d20));
```

MATRIX([[L1], [L2], [L3]]) = MATRIX([[1], [0], [0]])

```
> d21:=subs({X=X2,Y=Y2},op(2,d19)):
  op(natural)=map(simplify,op(d21));
```

MATRIX([[L1], [L2], [L3]]) = MATRIX([[0], [1], [0]])

NATURAL REFERENCE SYSTEMS   167

```
> d22:=subs({X=X3,Y=Y3},op(2,d19)):
  op(natural)=map(simplify,op(d22));
```

MATRIX([[L1], [L2], [L3]]) = MATRIX([[0], [0], [1]])

These nodal area coordinates are shown in Fig. F26.

One constructs similarly the volume coordinates in the case of a tetrahedron element, which has four vertices,

$$L_i = V_i/V, \quad i = 1, 2, 3, 4 \tag{F29.7}$$

where V is the tetrahedron volume and V_1 = volume $c423$, etc. (Fig. F27). Note that the point c is not a node. It can be shown that the volume coordinates of the nodes 1, 2, 3, and 4 are (1,0,0,0), (0,1,0,0), (0,0,1,0) and (0,0,0,1), respectively.

Below we present a program which computes the relations between the cartesian coordinates X, Y, Z and the volume coordinates L_1, L_2, L_3 and L_4:

```
> cartesian:=array([[1],[X],[Y],[Z]]);
```

cartesian := MATRIX([[1], [X], [Y], [Z]])

```
> natural:=array([[L1],[L2],[L3],[L4]]);
```

natural := MATRIX([[L1], [L2], [L3], [L4]])

```
> transform:=array([[1,1,1,1],[X1,X2,X3,X4],
  [Y1,Y2,Y3,Y4],[Z1,Z2,Z3,Z4]]);
```

transform := MATRIX([[1, 1, 1, 1], [X1, X2, X3, X4], [Y1, Y2, Y3, Y4], [Z1, Z2, Z3, Z4]])

```
> d25:=multiply(transform,natural):
> op(cartesian)=map(simplify,op(d25));
```

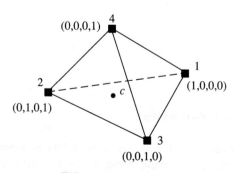

F27 Tetrahedron element

MATRIX([[1], [X], [Y], [Z]]) = MATRIX([[L1 + L2 + L3 + L4],
[X1*L1 + X2*L2 + X3*L3 + X4*L4], [Y1*L1 + Y2*L2 + Y3*L3 + Y4*L4],
[Z1*L1 + Z2*L2 + Z3*L3 + Z4*L4]])

```
> d27:=inverse(transform):
> d28:=op(natural)=multiply(d27,cartesian):
```

Now compute the volume coordinates of the nodes:

```
> d29:=subs({X=X1,Y=Y1,Z=Z1},op(2,d28)):
  op(natural)=map(simplify,op(d29));
```

MATRIX([[L1], [L2], [L3], [L4]]) = MATRIX([[1], [0], [0], [0]])

```
> d30:=subs({X=X2,Y=Y2,Z=Z2},op(2,d28)):
  op(natural)=map(simplify,op(d30));
```

MATRIX([[L1], [L2], [L3], [L4]]) = MATRIX([[0], [1], [0], [0]])

```
> d31:=subs({X=X3,Y=Y3,Z=Z3},op(2,d28)):
  op(natural)=map(simplify,op(d31));
```

MATRIX([[L1], [L2], [L3], [L4]]) = MATRIX([[0], [0], [1], [0]])

```
> d32:=subs({X=X4,Y=Y4,Z=Z4},op(2,d28)):
  op(natural)=map(simplify,op(d32));
```

MATRIX([[L1], [L2], [L3], [L4]]) = MATRIX([[0], [0], [0], [1]])

The natural coordinate systems may facilitate the evaluation of integrals. For example, using length coordinates,

$$\int_L L_1^a L_2^b \, dL = a!b!L/(a+b+1)! \tag{F29.8}$$

where L is the length between the adjacent nodes. Similarly, using area coordinates,

$$\int_S L_1^a L_2^b L_3^c \, dS = a!b!c!2S/(a+b+c+2) \tag{F29.9}$$

where S is the triangle area. In both cases the exponents must be positive integers.

F30. Serendipity coordinates

Another type of normalized dimensionless coordinates consists of the so-called serendipity coordinates, the name being originated from a Persian fairy tale and referring to a discovery by chance.

SERENDIPITY COORDINATES

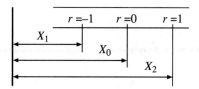

F28 Two-nodes lineal element

As Fig. F28 shows, the serendipity coordinate r is

$$r = 2(X - X_0)/(X_2 - X_1) \tag{F30.1}$$

where X_1 and X_2 are the nodal global coordinates and

$$X_0 = (X_1 - X_2)/2 \tag{F30.2}$$

Consequently, $-1 \leq r \leq 1$. The one-dimensional shape functions (F25.4)

$$d_1(x) = (x_2 - x)/(x_2 - x_1), \qquad d_2(x) = (x - x_1)/(x_2 - x_1) \tag{F30.3}$$

become

$$d_1(r) = (1 - r)/2, \qquad d_2(r) = (1 + r)/2 \tag{F30.4}$$

This type of one-dimensional coordinates has in fact been given by (F29.3) and (F29.4) In the case of a four-node rectangular element (Fig. F29), the two coordinates are

$$\xi = (X - X_0)/b, \qquad \eta = (Y - Y_0)/a \tag{F30.5}$$

with X_0 and Y_0 being the coordinates of the midpoint. The nodal serendipity coordinates are also shown in Fig. F29. The shape functions (F22.1)

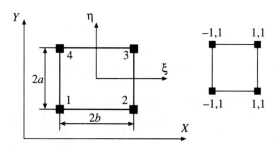

F29 Four-node rectangular element

$$d[1](x) := -1/4\frac{yb - xy + xa - ba}{ba}$$
$$d[2](x) := 1/4\frac{xa + ba - xy - yb}{ba}$$
$$d[3](x) := 1/4\frac{xa + ba + xy + yb}{ba} \quad \text{(F30.6)}$$
$$d[4](x) := -1/4\frac{-ba + xa - yb + xy}{ba}$$

take a simpler form:
$$d_1(\xi, \eta) = (1 - \xi)(1 - \eta)/4, \quad d_2(\xi, \eta) = (1 + \xi)(1 - \eta)/4,$$
$$d_3(\xi, \eta) = (1 + \xi)(1 + \eta)/4, \quad d_4(\xi, \eta) = (1 - \xi)(1 + \eta)/4, \quad \text{(F30.7)}$$

Integrals of interest may be conveniently evaluated in serendipity coordinates, for example:

$$\int_S d_i(x, y) d_j(x, y) \, ds = \int_{-1}^{1}\int_{-1}^{1} d_i(\xi, \eta) d_j(\xi, \eta) ab \, d\xi d\eta = S/18 \quad \text{(F30.8)}$$

which may be verified with the help of MAPLE.

F31. Concept of isoparametric elements

(MAPLE file vfem44)

In previous considerations the expression for the field variable, say, the deflection or temperature, has been first written as a polynomial with 'free' coefficients. Then by a proper procedure these coefficients have been eliminated from the analysis, so as to introduce instead nodal quantities. An alternative way is to use the nodal quantities from the beginning.

Consider a one-dimensional element with the two nodes, $L = -1$ and $L = 1$, respectively. The shape functions follow from (F29.3)

$$d_1(L) = (1 - L)/2, \quad d_2(L) = (1 + L)/2 \quad \text{(F31.1)}$$

and for their nodal values we get

$$d_1(L = -1) = 1, \quad d_1(L = 1) = 0,$$
$$d_2(L = -1) = 0, \quad d_2(L = 1) = 1 \quad \text{(F31.2)}$$

Note that (F31.2) is an illustration of the basic property of shape functions expressed by (F1.10).

If the field variable is u and the nodal vector is $\{u\}$,

$$\{u\} = [u_1, u_2]^T \quad \text{(F31.3)}$$

then one may obviously set

$$u = \{d(L)\}^T\{u\} = [(1 - L)/2]u_1 + [(1 + L)/2]u_2 \quad \text{(F31.4)}$$

CONCEPT OF ISOPARAMETRIC ELEMENTS   171

Thus, from the very beginning the field variable has been formulated in terms of its nodal vector.

Similarly, in case of a three-node lineal element, in the serendipity coordinate r the nodes have coordinates $r = -1$, $r = 1$ and $r = 0$. The shape functions are

$$d_1(r) = r(r-1)/2, \quad d_2(r) = r(r+1)/2, \quad d_3(r) = 1 - r^2 \qquad \text{(F31.5)}$$

and the nodal vector is

$$u = [u_1, u_2, u_3]^T \qquad \text{(F31.6)}$$

So, once again one may set

$$u = \{d(r)\}^T \{u\} \qquad \text{(F31.7)}$$

Now we turn to the transformation between the natural and cartesian coordinates, or, in more general terms, to the transformation of 'geometry'. Note that in case, for example, of the two-node element, we may set the following mapping:

$$X = \{d(L)\}^T \{X\} \qquad \text{(F31.8)}$$

where

$$\{X\} = [X_1, X_2]^T \qquad \text{(F31.9)}$$

Here use has been made of (F31.1) and (F31.2). The equation (F31.8) is merely (F29.3). If (F31.8) is combined with (F31.4), in other words, the geometry and the field variable both are given by the same transformation, the element is said to be *isoparametric*. The above results are of quite a general nature and may also be established for the plane and spatial elements.

As an application of this approach, consider an axial load finite element shown in Fig. F30.

In view of the 'smallness' of the finite element, the cross-sectional area S may be thought of as a linear function of the natural coordinate L (or the serendipity coordinate r). Using the coordinate L, which is given by (F29.3), get

$$S = \{d(L)\}^T \{S\} \qquad \text{(F31.10)}$$

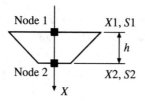

F30 Axial load element

with

$$\{S\} = [S_1, S_2]^T, \quad \{d(L)\} = [(1-L)/2, (1+L)/2]^T \quad (F31.11)$$

Here use has been made of (F29.3) and of the notations S_1 and S_2 for the nodal areas. The equation (F31.10) is a particular case of the above equation (F31.8) dealing with the 'geometry'.

On neglecting a possible buckling, the only field variable is the axial displacement $u(L)$, which is set, according to (F31.4), as

$$u = \{d(L)\}^T\{u\} = (1-L)/2 u_1 + (1+L)/2 u_2 \quad (F31.12)$$

In view of (F29.4), the strain ϵ is

$$\epsilon = du/dX = (du/dL)(dL/dX) = (2/h)\, du/dL \quad (F31.13)$$

which, because of (F31.12), takes a familiar matrix form

$$\epsilon = \{B\}^T\{u\} \quad (F31.14)$$

Here the strain–displacement matrix $\{B\}$ is

$$\{B\} = (1/h)[-1, 1]^T \quad (F31.15)$$

with $h = X_2 - X_1$. The strain energy U is

$$U = 1/2 \int_{X_1}^{X_2} ES\epsilon^2\, dx \quad (F31.16)$$

Taking into account that, in view of (F31.14),

$$\epsilon^2 = \epsilon\epsilon^T = \{u\}^T\{B\}\{B\}^T\{u\} \quad (F31.17)$$

and that $dX = h\,dL/2$, we get

$$U = 1/2\{u\}^T[K]\{u\} \quad (F31.18)$$

with the stiffness matrix $[K]$ given by

$$[K] = \int_{X_1}^{X_2} \{B\}ES\{B\}^T\, dX \quad (F31.19)$$

Note that the structure of (F31.19) is similar to that of the stiffness matrices obtained for the other cases, for example (F21.2). Taking into account that in this case $\{B\}$ is a constant and S is a function of L, (F31.19) may be written as

$$[K] = Eh/2\{B\}\{B\}^T \int_{-1}^{1} \{d(L)\}^T\{S\}\, dL \quad (F31.20)$$

where use has been made of (F31.10). The computation is therefore as follows:

```
> B:=evalm(array([[-1],[1]])/h):
> L:='L':d:=array([[(1-L)/2],[(1+L)/2]]):
> S:=multiply(transpose(d),array([[S1],[S2]])):
> int(S[1,1],L=-1..1):
> evalm(Eh/2*multiply(B,transpose(B))):
> K:=evalm("""*");
```

$$K := \begin{bmatrix} 1/2\dfrac{(S1+S2)Eh}{h^2} & -1/2\dfrac{(S1+S2)Eh}{h^2} \\ -1/2\dfrac{(S1+S2)Eh}{h^2} & 1/2\dfrac{(S1+S2)Eh}{h^2} \end{bmatrix}$$

where $\mathtt{Eh} = Eh$, $\mathtt{S1} = S_1$, etc.

In general, the approximations used for the geometry transformation and for the field variable should not be necessarily of the same order. For example, defining the geometry with the help of the two-parametric transformation (F31.8) and (F31.9), and the approximation for the field variable as the three-parametric mapping (F31.5), (F31.6) and (F31.7), one gets a so-called *subparametric* element. For the *superparametric* element the order of the geometry transformation is higher than that of the field approximation.

F32. Three-node isoparametric bar element

(MAPLE file vfem45)

A three-node element follows from the shape functions (F31.5):

$$d_1(r) = r(r-1)/2, \qquad d_2(r) = r(r+1)/2, \qquad d_3(r) = 1 - r^2 \qquad \text{(F32.1)}$$

If the element length is h (see Fig. F31), then the nodal coordinate X_3 may be anywhere inside the element.

According to the concept of isoparametric element, set for the geometry

$$X = [d_1, d_2, d_3][X_1, X_2, X_3]^T \qquad \text{(F32.2)}$$

and for the axial displacement u

$$u = [d_1, d_2, d_3][u_1, u_2, u_3]^T \qquad \text{(F32.3)}$$

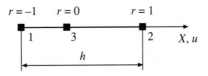

F31 Three-node bar element

174 FINITE ELEMENT METHODS

The strain is

$$\epsilon = du/dX = (du/dr)(dr/dX) \tag{F32.4}$$

which leads to the strain–displacement matrix $[B]$ appearing in the expression for $[K]$ (see, for instance, (F31.19)).

Turning to computations, formulate the shape functions (F32.1):

```
> with(linalg,multiply,transpose):
> d:=array(1..1,1..3);
```

$$d := \text{array}(1..1, 1..3, [\,])$$

```
> d[1,1]:=(-r+r^2)/2;
```

$$d[1,1] := -1/2\, r + 1/2\, r^2$$

```
> d[1,2]:=(r+r^2)/2;
```

$$d[1,2] := 1/2\, r + 1/2\, r^2$$

```
> d[1,3]:=1-r^2;
```

$$d[1,3] := 1 - r^2$$

and (F32.2) and (F32.3)

```
> Xn:=array([[X1],[X2],[X3]]):
> un:=array([[u1],[u2],[u3]]):
> X:=multiply(d,Xn);
```

$$X := [-1/2\, X1\, r + 1/2\, X1\, r^2 + 1/2\, X2\, r + 1/2\, X2\, r^2 + X3 - X3\, r^2]$$

```
> u:=multiply(d,un);
```

$$u := [-1/2\, u1\, r + 1/2\, u1\, r^2 + 1/2\, u2\, r + 1/2\, u2\, r^2 + u3 - u3\, r^2]$$

Instead of computing dr/dX appearing in (F32.4), compute the Jacobian $dX/dr = J$:

```
> J:=diff(X[1,1],r);
```

$$J := -1/2\, X1 + X1\, r + 1/2\, X2 + X2\, r - 2\, X3\, r$$

Now, in view of (F32.3) and (F32.4), $[B] = $ B is

```
> B:=map(diff,op(d),r)/J;
```

$$B := \begin{bmatrix} -1/2 + r, & 1/2 + r, & -2r \\ -1/2\,X1 + X1\,r + 1/2\,X2 + X2\,r - 2\,X3\,r \end{bmatrix}$$

The stiffness matrix $[K]$ is similar to (F31.19):

$$[K] = \int_{X_1}^{X_2} \{B\}ES\{B\}^T \, dX = \int_{-1}^{1} \{B\}ES\{B\}^T J \, dr \tag{F32.5}$$

F33. Two-dimensional isoparametric element

An additional advantage of isoparametric elements is their ability to better approximate the boundaries. In general, the possibility of using a variety of elements rather than those having trivial geometries further enhances the efficacy of the technique.

Consider a theory of isoparametric four-node (bilinear) elements. Introducing the serendipity coordinates ξ and η (Section F30), set for the geometry

$$X = d_i(\xi, \eta) X_i, \qquad Y = d_i(\xi, \eta) Y_i \tag{F33.1}$$

with the summation with respect to the repeated subscript i from 1 to 4.

Similarly, for the field variable, say, temperature T

$$T = d_i(\xi, \eta) T_i \tag{F33.2}$$

Here X_i, Y_i are the nodal coordinates, T_i the nodal temperature and $d_i(\xi, \eta)$ the shape functions given by (F30.7)

$$\begin{aligned} d_1(\xi, \eta) &= (1 - \xi)(1 - \eta)/4, & d_2(\xi, \eta) &= (1 + \xi)(1 - \eta)/4, \\ d_3(\xi, \eta) &= (1 + \xi)(1 + \eta)/4, & d_4(\xi, \eta) &= (1 - \xi)(1 + \eta)/4, \end{aligned} \tag{F33.3}$$

The above relations (F33.1) and (F33.3) deal with the mapping between the XY-plane and $\xi\eta$-plane, as shown in Fig. F32. The mapping is unique once the nodal coordinates X_i and Y_i are given. The $\xi\eta$-element may be considered as a parent one. Note that, in general, one should avoid the use of severely distorted elements.

176 FINITE ELEMENT METHODS

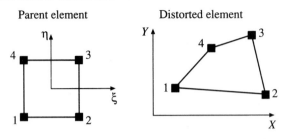

F32 Mapping of bilinear element

In view of (F24.16), the relevant 'stiffness' matrix may be written as

$$[K] = \int\int_S [B]^T[H][B] t \, dx \, dy \tag{F33.4}$$

which must be modified so as to incorporate the serendipity coordinates. In the above $[H]$ is the conductivity matrix, which may reduce to a thermal conductivity k_0 in the isotropic case, and t is the thickness.

Changing the variables, get

$$[K] = \int_{-1}^{1}\int_{-1}^{1} [B]^T[H][B] tJ \, d\xi \, d\eta \tag{F33.5}$$

Below we evaluate $[B]$ and the Jacobian $J = \det[J]$ in the $\xi\eta$-coordinate system. The Jacobian matrix $[J]$ follows from the relation between the derivatives of T with respect to X and Y and those with respect to ξ and η:

$$[T_{,\xi} \quad T_{,\eta}]^T = [J][T_{,X} \quad T_{,Y}]^T \tag{F33.6}$$

or in a more explicit form

$$\begin{aligned} T_{,\xi} &= T_{,X} X_{,\xi} + T_{,Y} Y_{,\xi} \\ T_{,\eta} &= T_{,X} X_{,\eta} + T_{,Y} Y_{,\eta} \end{aligned} \tag{F33.7}$$

This specifies the entries of $[J]$ as

$$\begin{aligned} J[1,1] &= X_{,\xi} & J[1,2] &= Y_{,\xi} \\ J[2,1] &= X_{,\eta} & J[2,2] &= Y_{,\eta} \end{aligned} \tag{F33.8}$$

which may be explicitly evaluated with the help of (F33.1) and (F33.3). Hence, the Jacobian $J = \det [J]$ is

$$J = X_{,\xi} Y_{,\eta} - X_{,\eta} Y_{,\xi} \tag{F33.9}$$

As to $[B]$ appearing in (F33.5), observe from (F33.6) that

$$[T_{,X} \quad T_{,Y}]^T = [J]^{-1}[T_{,\xi} \quad T_{,\eta}]^T \tag{F33.10}$$

and from the definition of $[B]$ in case of heat conduction (see F24.15) that

$$[T_{,X} \quad T_{,Y}]^T = [B]\{T\} \tag{F33.11}$$

In turn, from (F33.2),

$$[T_{,\xi} \quad T_{,\eta}]^T = [d(\xi,\eta)/d\xi, \quad d(\xi,\eta)/d\eta]^T \{T\} \tag{F33.12}$$

with $d(\xi,\eta) = d_1(\xi,\eta), d_2(\xi,\eta), d_3(\xi,\eta), d_4(\xi,\eta)$. This eventually yields

$$[B] = [J]^{-1}[d(\xi,\eta)/d\xi, \quad d(\xi,\eta)/d\eta]^T \tag{F33.13}$$

Substituting (F33.13) and (F33.9) in (F33.5) leads to the 'stiffness' matrix $[K]$. In the next section we program the above relations.

F34. Programming for an isoparametric element

(MAPLE file vfem46)

First, program the relevant shape functions (F33.3)

$$\begin{aligned} d_1(\xi,\eta) &= (1-\xi)(1-\eta)/4, & d_2(\xi,\eta) &= (1+\xi)(1-\eta)/4, \\ d_3(\xi,\eta) &= (1+\xi)(1+\eta)/4, & d_4(\xi,\eta) &= (1-\xi)(1+\eta)/4, \end{aligned} \tag{F34.1}$$

noting that these may be written as

$$d_i(\xi,\eta) = (1+\xi\xi_i)(1+\eta\eta_i)/4, \quad i = 1, 2, 3, 4 \tag{F34.2}$$

with $\xi_i = -1, 1, 1, -1$ and $\eta_i = -1, -1, 1, 1$. Below $\xi = e$ and $\eta = h$,

```
> ce:=-1,1,1,-1;
    ce := -1, 1, 1, -1
> ch:=-1,-1,1,1;
    ch := -1, -1, 1, 1
> for i to 4 do
> d[i]:=(1+e*ce[i])*(1+h*ch[i])/4 od;i:='i':
    d[1] := 1/4 (1 - e) (1 - h)
    d[2] := 1/4 (1 + e) (1 - h)
    d[3] := 1/4 (1 + e) (1 + h)
    d[4] := 1/4 (1 - e) (1 + h)
```

178 FINITE ELEMENT METHODS

According to (F33.1) and (F33.2), the geometry transformation and the approximation for T are $(X_i = \text{X}[i], Y_i = \text{Y}[i], X = \text{Xx}, Y = \text{Yy}$ and $T = \text{Tt})$

```
>Xx:=sum(d[i]*X[i],i=1..4);
```

$\quad$ Xx := 1/4 (1 − e) (1 − h) X[1] + 1/4 (1 + e) (1 − h) X[2]
$\quad\quad$ + 1/4 (1 + e) (1 + h) X[3] + 1/4 (1 − e) (1 + h) X[4]

```
>Yy:=sum(d[i]*Y[i],i=1..4);
```

$\quad$ Yy := 1/4 (1 − e) (1 − h) Y[1] + 1/4 (1 + e) (1 − h) Y[2]
$\quad\quad$ + 1/4 (1 + e) (1 + h) Y[3] + 1/4 (1 − e) (1 + h) Y[4]

```
>Tt:=sum(d[i]*T[i],i=1..4);
```

$\quad$ Tt := 1/4 (1 − e) (1 − h) T[1] + 1/4 (1 + e) (1 − h) T[2]
$\quad\quad$ + 1/4 (1 + e) (1 + h) T[3] + 1/4 (1 − e) (1 + h) T[4]

Next, state the Jacobian matrix $[J] = \text{Jm}$ according to (F33.8) and compute its Jacobian J

```
>Jm:=array([[diff(Xx,e),diff(Yy,e)],[diff(Xx,h),
 diff(Yy,h)]]);
```

Jm :=
[− 1/4 (1 − h) X[1] + 1/4 (1 − h) X[2] + 1/4 (1 + h) X[3] − 1/4 (1 + h) X[4],
$\quad$ − 1/4 (1 − h) Y[1] + 1/4 (1 − h) Y[2] + 1/4 (1 + h) Y[3] − 1/4 (1 + h) Y[4]]
[− 1/4 (1 − e) X[1] − 1/4 (1 + e) X[2] + 1/4 (1 + e) X[3] + 1/4 (1 − e) X[4],
$\quad$ − 1/4 (1 − e) Y[1] − 1/4 (1 + e) Y[2] + 1/4 (1 + e) Y[3] + 1/4 (1 − e) Y[4]]

```
>with(linalg,det,inverse,multiply);
 [det, inverse, multiply]
>J:=det(Jm):
```

As to $[B]$, compute the derivatives of the shape functions according to (F33.12):

```
>for i to 4 do n[i]:=diff(d[i],e):m[i]:=diff(d[i],h):
 od:i:='i':
>array([[n[1],n[2],n[3],n[4]],[m[1],m[2],m[3],m[4]]]):
```

and then $[B] = \text{B}$ according to (F33.13):

```
>B:=multiply(inverse(Jm),"):
```

where the display of a complicated expression has been prevented.

So far we carried out the computation in a general symbolic way. As a particular case, let the nodal coordinates be

> nodes:=[X[1]=1,Y[1]=1,X[2]=4,Y[2]=2,X[3]=3,Y[3]=4,
  X[4]=1,Y[4]=4];

  nodes := [X[1] = 1, Y[1] = 1, X[2] = 4, Y[2] = 2, X[3] = 3, Y[3] = 4, X[4] = 1, Y[4] = 4]

Then the above quantities are

> Jm:=subs(nodes,op(Jm));

  Jm := MATRIX([[5/4 − 1/4*h, 1/4 − 1/4*h], [− 1/4 − 1/4*e, 5/4 − 1/4*e]])

> J:=subs(nodes,J);

  J := − 3/8*h − 1/4*e + 13/8

> B:=subs(nodes,op(B));

  B := MATRIX([[(2 − 2*h)/(3*h + 2*e − 13), −(3 − 3*h)/(3*h + 2*e − 13),
  −(2e + 3*h)/(3*h + 2*e − 13), (− e + 2*h + 3)/(3*h + 2*e − 13)],
  [−(2*e − 3 + h)/(3*h + 2*e − 13), (2 + 2*e)/(3*h + 2*e − 13),
  (− 3*e − 3)/(3*h + 2*e − 13), −(2 − 3*e − h)/(3*h + 2*e − 13)]])

Observe that $[B]$ contains ξ and η in both the denominators and numerators of its entries, which justifies a resort to numerical integrations while calculating $[K]$ via (F33.5):

$$[K] = \int_{-1}^{1} \int_{-1}^{1} [B]^T k_0 [B] t J \, d\xi \, d\eta \qquad (F34.3)$$

F35. Special elements

(MAPLE file vfem45)

Functions of interest may show a singular behaviour, an example of which is a stress in the crack vicinity. Such a stress varies as $x^{-1/2}$, where x is the distance from the crack tip. Problems of this type may be studied with the help of conventional finite elements on condition that the mesh is sufficiently fine. An

180   FINITE ELEMENT METHODS

alternative is to use special elements, which would exhibit a needed singularity. For example, the three-node isoparametric element of Section F32 may be so modified as to show the singularity $x^{-1/2}$.

Continuing the program of Section F32, redefine the nodal coordinates (see Fig. F31, Section F32) so as to place the third coordinate at $h/4$,

```
> nodes:=[X1=0,X2=h,X3=h/4]:
```

and compute the 'geometry' $X(r)$ by

```
> subs(nodes,op(X));
```

$$[1/2\,h\,r + 1/4\,h\,r^2 + 1/4\,h]$$

Solving this for r, we get

```
> s:=solve(X-"[1,1],r);
```

$$s := -2\,\frac{1/2\,h + h^{1/2}\,X^{1/2}}{h},\ -2\,\frac{1/2\,h - h^{1/2}\,X^{1/2}}{h}$$

The root of interest here is $r = 2(X/h)^{1/2} - 1$. Recompute $[B] = $ B for this result:

```
> B:=evalm(map(diff,op(d),r)/J):
> subs(nodes,op(B)):B:=subs(r=2*(X/h)^(1/2)-1,"):
> B:=map(simplify,op(B));
```

$$B := \left[-1/2\,\frac{3h^{1/2} - 4X^{1/2}}{h\,X^{1/2}}\quad -1/2\,\frac{h^{1/2} - 4X^{1/2}}{h\,X^{1/2}}\quad -2\,\frac{-2X^{1/2} + h^{1/2}}{h\,X^{1/2}} \right]$$

which displays the desirable singularity $x^{-1/2}$. Since the stress is

$$\sigma = E[B]\{u\} \tag{F35.1}$$

it displays this singularity too.

Another 'non-standard' example is the so-called infinite element, which is convenient for unbounded or very large domains. One of the approaches is to use usual shape functions to represent a field variable, but for the 'geometry' resort to shape functions, which grow in an unlimited way for large distances from the origin.

As an example, consider again the three-node element shown in Fig. F33.

Fig. F33a shows the element using the local normalized coordinate r and Fig. F33b using the global coordinate X. Point 0 is a pole, the meaning of which should become clear later. The shape functions for the geometry are

$$d_1^{\text{inf}}(r) = -2r/(1-r), \qquad d_2^{\text{inf}}(r) = (1+r)/(1-r) \tag{F35.2}$$

SPECIAL ELEMENTS   181

(a)

(b)
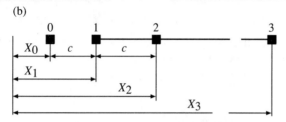

F33 Illustrating infinite element

and the coordinate transformation is

$$X = [d_1^{\text{inf}}, d_2^{\text{inf}}][X_1, X_2]^T \tag{F35.3}$$

or in the MAPLE code ($\text{dinf} = [d_1^{\text{inf}}, d_2^{\text{inf}}]^T$)

```
> dinf:=array([[-2*r/(1-r)],[(1+r)/(1-r)]]);
```

$$\text{dinf} := \begin{bmatrix} -2r/(1-r) \\ (1+r)/(1-r) \end{bmatrix}$$

```
> X:=dinf[1,1]*X1+dinf[2,1]*X2;
```

$$X := -2\frac{r\,X1}{1-r} + \frac{(1+r)\,X2}{1-r}$$

To find the global coordinates of the nodes $r = -1$ and $r = 0$, type

```
> subs(r=-1,X):simplify(");
    X1
> subs(r=0,X):simplify(");
    X2
```

As to the value of the third coordinate $r = 1$, get

```
> X3:=limit(X,r=1,left);
    X3 := signum(-2 X1 + 2 X2) infinity
```

Since $X_2 > X_1$, this means that $X_3 \to \infty$, as needed.

A simple choice of the shape functions for a field variable is

$$d_1(r) = (1-r)/2, \qquad d_2(r) = (1+r)/2 \tag{F35.4}$$

182 FINITE ELEMENT METHODS

This yields
$$T = d_1 T_1 + d_2 T_2 \tag{F35.5}$$

The strain–displacement B-matrix is
$$[B] = [d_{1,r} \; d_{2,r}]/J, \quad J = dX/dr = d_1^{\inf}(r)_{,r} X_1 + d_2^{\inf}(r)_{,r} X_2 \tag{F35.6}$$

To understand the meaning of the pole 0, compute first the explicit expression for T according to (F35.5)

```
> d:=array([[(1-r)/2],[(1+r)/2]]);
```
$$d := \begin{bmatrix} 1/2 - 1/2r \\ 1/2 + 1/2r \end{bmatrix}$$

```
> T:=multiply(transpose(d),array([[T1],[T2]]));
```
$$T := [\,1/2\ T1 - 1/2\ T1\ r + 1/2\ T2 + 1/2\ T2\ r\,] \tag{F35.7}$$

Referring to Fig. F33 and introducing the coordinate $Y = X - X_0$ with the origin at 0, find the function $r = r(Y)$ by

```
> solve(X0+Y-X,r):subs(X2=X1+c,"):
> r:=subs(X1=X0+c,");
```
$$r := -\frac{-Y + 2c}{Y}$$

With this expression for r, (F35.7) shows that the point 0 ($Y = 0$) is a singular point of the field variable T.

F36. Constraints and Lagrange multipliers

(MAPLE file vfem47)

As noted in Section D17, the presence of various constraints is typical of engineering problems and their account may be handled with the help of Lagrange multipliers. Consider a composite bar, shown in Fig. F34. Besides the condition for the axial displacement $U_1^a = 0$ imposed at $x = 0$ (node 1), set the constraint $U_2^a = U_3^a$ and find the value of U_2^a as it depends on the force P. (Clearly, the correct result should indicate that the element 2 behaves as a rigid body to comply with the constraint $U_2^a = U_3^a$.)

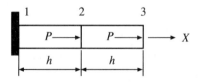

F34 Two bar elements

CONSTRAINTS AND LAGRANGE MULTIPLIERS   183

Generalizing the results for the elemental strain energy, obtained in Sections F2 and F3, one may set for the potential energy of the assembly

$$U = 1/2\{W\}_a^T[K]_a\{W\}_a - \{W\}_a\{R\}_a \qquad (F36.1)$$

where the second term is the work done by the external forces. Instead of looking for a minimum of (F36.1) subjected to the above constraint, introduce the modified functional to be extremized *unconditionally*:

$$F = 1/2\{W\}_a^T[K]_a\{W\}_a - \{W\}_a\{R\}_a + \lambda(U_2^a - U_3^a) \qquad (F36.2)$$

where $\{W\}_a = [U_1^a, U_2^a, U_3^a]$ with $U_1^a = 0$ and λ is the Lagrange multiplier. Now the stationarity of F implies the equations

$$\partial F/\partial \lambda = 0, \qquad \partial F/\partial U_2^a = 0, \qquad \partial F/\partial U_3^a = 0 \qquad (F36.3)$$

the solution of which should provide U_2^a, U_3^a and λ. Going over to symbolic computations (we drop the index 'a'), state the elemental stiffness matrix k1, as derived in Section F3,

```
> k1:=evalm(EA/h*array([[1,-1],[-1,1]]));
```

$$k1 := \begin{bmatrix} \dfrac{EA}{h} & -\dfrac{EA}{h} \\ -\dfrac{EA}{h} & \dfrac{EA}{h} \end{bmatrix}$$

The connectivity of the element 2 is given below by c2, the expanded matrices and the assembly stiffness matrix Ka:

```
> c2:=2,3:
> K1ex:=array(1..3,1..3,sparse):K2ex:=array(1..3,1..3,
   sparse):
> for i to 2 do for j to 2 do K1ex[i,j]:=k1[i,j] od od:
> for i to 2 do for j to 2 do K2ex[c2[i],c2[j]]:=k1[i,j] od od:
> Ka:=evalm(K1ex+K2ex);
```

$$Ka := \begin{bmatrix} \dfrac{EA}{h} & -\dfrac{EA}{h} & 0 \\ -\dfrac{EA}{h} & 2\dfrac{EA}{h} & -\dfrac{EA}{h} \\ 0 & -\dfrac{EA}{h} & \dfrac{EA}{h} \end{bmatrix}$$

Next, formulate the displacement and external forces and the functional F given by (F36.2):

```
> Ra:=array([[R1],[P],[P]]):
> Wa:=array([[U1],[U2],[U3]]):
```

```
> with(linalg,multiply,transpose):
> multiply(transpose(Wa),Ka):evalm(multiply(",Wa)/2):
> F:=evalm("-multiply(transpose(Wa),Ra)+array([[L*U2-
    L*U3]]));
```

$$F := [1/2 \frac{EA(U1^2 - 2\,U1\,U2 + 2\,U2^2 - 2\,U2\,U3 + U3^2)}{h} - U1\,R1$$
$$- U2\,P - U3\,P + L\,U2 - L\,U3]$$

where $L = \lambda$. It remains to compute the stationarity conditions (F36.3), substitute $U_1^a = 0$ and solve for U_2^a, U_3^a and $L = \lambda$:

```
> eq1:=diff(F[1,1],U2);
```

$$eq1 := 1/2 \frac{EA(-2\,U1 + 4\,U2 - 2\,U3)}{h} - P + L$$

```
> eq2:=diff(F[1,1],U3);
```

$$eq2 := 1/2 \frac{EA(-2\,U2 + 2\,U3)}{h} - P - L$$

```
> eq3:=diff(F[1,1],L);
```

$$eq3 := U2 - U3$$

```
> eq1:=subs(U1=0,eq1);
```

$$eq1 := 1/2 \frac{EA(4\,U2 - 2\,U3)}{h} - P + L$$

```
> solve({eq1,eq2,eq3},{U2,U3,L});
```

$$\{U3 = 2\frac{P\,h}{EA}, L = -P, U2 = 2\frac{P\,h}{EA}\}$$

As expected, these results indicate that the element 2 behaves as a rigid one.

F37. Constraints and penalty parameters

(MAPLE file vfem47)

The above method of Lagrange multipliers reduces a constrained variational problem to the unconstrained problem but leads to a larger number of unknowns. An alternative approach, which may apply to a constrained minimization problem, is the *penalty* method. To introduce the idea, consider

CONSTRAINTS AND PENALTY PARAMETERS

the minimum of the functional of the static potential energy

$$U = \int_V u\, dv - D \qquad (F37.1)$$

where u is the strain energy density and D the work of external forces, and assume that the kinematic constraint is

$$f(W_1, W_2, W_3) = 0 \qquad (F37.2)$$

This constrained minimization problem may be stated approximately as the unconstrained one for the modified functional

$$U_p = \int_V u\, dv - D + \gamma/2 \int_V f^2(W_1, W_2, W_3)\, dv \qquad (F37.3)$$

where γ is a prescribed *large* (positive) penalty parameter. If $f(W_1, W_2, W_3) = 0$, the last term in (F37.3) vanishes. Under the above conditions on γ, the very form of (F37.3) suggests that enforcing its minimum implies also compliance with the constraint (F37.2).

As an illustration, consider again the problem of the two-element bar treated in Section F36 (Fig. F34). The functional (F37.3) takes the form

$$U_p = 1/2\{W\}_a^T [K]_a \{W\}_a - \{W\}_a \{R\}_a + \gamma/2(U_2^a - U_3^a)^2 (EA/h) \qquad (F37.4)$$

where the multiplier EA/h enables one to treat the penalty parameter γ as dimensionless. The minimum of (F37.4) implies

$$\partial U_p/\partial U_2^a = 0, \qquad \partial U_p/\partial U_3^a = 0 \qquad (F37.5)$$

Going over to MAPLE, as a sequel to the previous program, state the constraint $U_2^a - U_3^a = 0$ as the array a and then (F37.4)

```
> a:=array([[U2-U3]]);
    a := [U2 − U3]
> c:=evalm(gamma/2*EA/h*multiply(a,transpose(a))):
> Up:=evalm(s-multiply(transpose(Wa),Ra)+c):
  Up:=map(simplify, op(Up))
```

$$Up := [1/2\,(2\,EA\,U2^2 - 2\,EA\,U2\,U3 + EA\,U3^2 - 2\,U2\,P\,h - 2\,U3\,P\,h$$
$$+ gamma\,EA\,U2^2 - 2\,gamma\,EA\,U2\,U3 + gamma\,EA\,U3^2)/h]$$

Next, state the boundary condition and (F37.5),

```
> U1:=0:
> eq1:=diff(Up[1,1],U2);
    eq1 :=
```
$$1/2\,\frac{4\,U2\,EA - 2\,U3\,EA - 2\,P\,h + 2\,gamma\,EA\,U2 - 2\,gamma\,EA\,U3}{h}$$

```
> eq2:=diff(Up[1,1],U3);
```

$$eq2 := \frac{1}{2}\frac{-2\,U2\,EA + 2\,U3\,EA - 2\,P\,h - 2\,\text{gamma}\,EA\,U2 + 2\,\text{gamma}\,EA\,U3}{h}$$

Resolving these for the unknowns, we get

```
> z:=solve({eq1,eq2},{U2,U3});
```

$$z := \{U2 = 2\frac{P\,h}{EA}, U3 = \frac{P\,h(3 + 2\,\text{gamma})}{EA(1 + \text{gamma})}\}$$

The penalty parameter γ should be as large as possible, so investigate the limit

```
> assign(z):
> limit(U3,gamma=infinity);
```

$$2\frac{P\,h}{EA}$$

It is seen that the results are identical to those obtained earlier with the help of Lagrange multipliers (Section F36). In numerical calculations the penalty parameters should be prescribed sufficiently large, without causing, at the same time, computational difficulties.

F38. Accuracy, convergence and related subjects

The finite element method is an essentially approximate technique and the results obtained with its help need a thorough inspection. For example, though the equilibrium (motion) equations are satisfied for the nodes, they may be violated within the elements because of the approximate trial functions. The obvious way to improve the accuracy is to refine the mesh and/or apply a higher-order shape function. Depending on the particular problem at hand, there may or may not be convergence to the exact solution.

The convergence requirements include a capability to support the state of rigid-body motion with zero strain. Since the element may be thought of as situated at any location within the domain, it must be capable of supporting a variety of stress states. Among these, the above state is a trivial one. The analysis of eigenvalues and eigenvectors of the stiffness matrix, as shown in Section F5, may provide a useful insight into the element behaviour.

The absence of vanishing eigenvalues may indicate that the element does not support a rigid-body motion nor a 'mechanism' mode. If an eigenvalue shows a dependence on the choice of coordinate axes, the element is not 'isotropic', which may place limitations on its use.

Another useful criterion is the capability to support a constant strain under appropriate nodal displacement. Indeed, it is intuitively clear that with the diminishing size of elements, they should approach a state of constant strain.

An element of a finite size should attain this property. In fact, it can be shown that this criterion embodies the above one of a rigid body displacement.

Compared to the classical direct methods, such as the Rayleigh–Ritz technique, the finite element method introduces a new dimension, which is a spatial discretization of the domain (structure). Consequently, the compatibility commented on in Section F8 requires a special discussion.

Note that the continuity is satisfied at the nodes and within the element, provided the approximation for the field variable is continuous. Ensuring the correspondence between the order of approximating polynomial and the number of nodes, one may impose continuity along the interelement boundaries too.

Nevertheless, this does not ensure consistent field for the derivatives. Recall that a field is said to have C^n continuity if its derivatives through order n are continuous. The C^0 elements may be used if, say, the displacement is of primary interest, while C^1 elements may be useful for finding the displacement and the slope.

Consider, for example, a plate element shown in Fig. F19 (Section F18) and the corresponding deflection w given by (F19.1):

$$w(x,y) = ao_1 + ao_2 x + ao_3 y + ao_4 x^2 + ao_5 xy + ao_6 y^2 + ao_7 x^3 + ao_8 x^2 y$$
$$+ ao_9 xy^2 + ao_{10} y^3 + ao_{11} x^3 y + ao_{12} xy^3 \tag{F38.1}$$

Then along the line 1–4, y = constant and w may be set as

$$w = N_0 + N_1 x + N_2 x^2 + N_3 x^3 \tag{F38.2}$$

while along the line 1–2 we similarly get

$$w = M_0 + M_1 y + M_2 y^2 + M_3 y^3 \tag{F38.3}$$

Here N_i and M_i, $i = 1, 2, 3, 4$ are constant coefficients, depending on the nodal vector $\{w\}$. It becomes clear that, for example, the mixed second derivatives at the node 1, $w_{,xy}$, namely those computed from (F38.2) and those computed from (F38.3), are not necessarily the same. This is an obvious violation of compatibility. Such elements are referred to as incompatible (non-conforming).

If the displacement suffers a discontinuity at an interface, then strains are infinite there and the process of integration in the element-by-element way is mathematically incorrect. Nevertheless, in general, the incompatibility should not be viewed as a fatal drawback. This is particularly true if the analysis is confined to the overall, not local, response. Allowing for discontinuities at the limited number of points (surfaces) may render a structure 'softer'. This, in turn, may reduce an error induced by resorting to direct methods, which usually overestimate the stiffness. Note that in this case the bounding feature of the result is no longer valid.

188 FINITE ELEMENT METHODS

The question arises about the convergence of results obtained through incompatible elements to the exact ones. The so-called *patch test* provides a condition for such convergence. Loosely speaking, one considers a small number of elements connected in, at least, one node. If the boundary nodes are loaded so as to allow for a state of constant strain and the 'patch' indeed supports this state, then the test is passed and with the mesh refinement the result would reproduce the exact one.

F39. More on element deficiency

Some of the possible element 'disorders' were mentioned in Section F38. The notion of *aliasing* enables one to gain a better insight into element performance. The alias or *substitute function* is a function which interpolates the nodal displacements within an element when these are uniquely specified for a particular problem. The equation governing the alias $u^a(x, y)$ for, say a two-dimensional problem, is given by the familiar expression

$$u^a(x, y) = \{d(x, y)\}^T \{u\} \tag{F39.1}$$

where $\{d(x, y)\}$ is the interpolation matrix and $\{u\}$ the nodal vector. Given $\{u\}$ for a particular problem at hand, one finds the alias from (F39.1).

To this end, consider the four-node rectangular element introduced in Section F22 and shown in Fig. F35.

Assume that this element applies to investigation of the in-plane bending by the moments acting on its vertical sides. Then the plane stress theory provides the following displacement field:

$$u = xy, \qquad v = -x^2/2 \tag{F39.2}$$

where u and v are the displacements along the x- and y-axis, respectively.

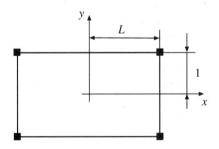

F35 Four-node (membrane) element

Accordingly, the strains and the stresses are

$$\epsilon_x = y, \qquad \epsilon_y = 0, \qquad \gamma_{xy} = 0$$
$$\sigma_x = Ey/(1 - v^2), \qquad \sigma_y = vEy/(1 - v^2), \qquad \tau_{xy} = 0 \qquad \text{(F39.3)}$$

Finding $\{u\}$ from (F39.2) and Fig. F35 and using the interpolation matrix from Section F22 one specifies $u^a(x, y)$. This is left to the reader as an exercise, as in this simple case it may be suggested in a straightforward way that

$$u^a(x, y) = xy \qquad \text{(F39.4)}$$

Similar considerations give

$$v^a(x, y) = -L^2/2 \qquad \text{(F39.5)}$$

Equations (F39.4) and (F39.5) yield the following strains and stresses:

$$\epsilon_x = y, \qquad \epsilon_y \bar{0}, \qquad \gamma_{xy} = x$$
$$\sigma_x = Ey/(1 - v^2), \qquad \sigma_y = vEy/(1 - v^2), \qquad \tau_{xy} = Ex/[2(1 + v)] \qquad \text{(F39.6)}$$

Note the difference in the shear strain and stress as compared with (F39.3). In order to estimate the deviation of the alias from the elasticity solution (F39.2) compute the strain energy for both cases. Equations (F39.4) and (F39.5) yield

$$U^a = 2EL[1 + (1 - v)L^2/2]/[3(1 - v^2)] \qquad \text{(F39.7)}$$

while the elasticity solution yields

$$U^{\text{elast}} = 2EL/[3(1 - v^2)] \qquad \text{(F39.8)}$$

Thus the ratio U^a/U^{elast}, which also governs the ratio of the element stiffness to the correct stiffness, is just the factor $[1 + (1 - v)L^2/2]$. It is seen that for a slender element, say $L = 5$, the error is large. With increasing L the element stiffness becomes extremely high. This phenomenon illustrates a particular case of *locking*, the so-called *shear locking*.

It is an easy matter to track the reason for this failure of the element performance to the spurious shear strain appearing in (F39.6), namely $\gamma_{xy} = x$ instead of $\gamma_{xy} = 0$, as (F39.3) shows. One way to remedy this is artificially to set $\gamma_{xy} = 0$ while integrating the strain energy density, which is known as *selective under-integration*. Though this technique obviously has a deficiency of its own, it is widely used.

F40. Symbolic database

(MAPLE file vfem48)

With the help of MAPLE one may create his own database containing the relevant quantities in a symbolic form. For example, the so-called 'strain–displacement' matrix B appears as a basic term in construction of the elemental

190 FINITE ELEMENT METHODS

characteristic (stiffness) matrix (see, for example, F18.20, F21.2, F32.5 and F33.4), regardless of the element shape and form of the approximation.

Below we deal with programs which derive in an automated way the B-matrix (provided the element topology and shape functions are given) and may therefore be used for the above purpose. Consider, for example, the isoparametric quadrilateral element in the plane-stress problem. The strain–displacement relation in cartesian coordinates X and Y is

$$\{e\} = \begin{bmatrix} 1 & 0 & 0 & 0 \\ 0 & 0 & 0 & 1 \\ 0 & 1 & 1 & 0 \end{bmatrix} \{c\} \qquad \text{(F40.1)}$$

where the strain $\{\epsilon\}$ is $[\epsilon_X \ \epsilon_Y \ \gamma_{XY}]^T$ and the displacement $\{c\}$ is $[U_{,X} \ U_{,Y} \ V_{,X} \ V_{,Y}]^T$. Next, the transformation from the displacement in the $\xi\eta$-coordinates given by $\{g\} = [U_{,\xi} \ U_{,\eta} \ V_{,\xi} \ V_{,\eta}]^T$ to the above $\{c\}$ is given by the 4×4 matrix

$$\begin{bmatrix} J^{-1} & \\ & J^{-1} \end{bmatrix} \qquad \text{(F40.2)}$$

with J being the 2×2 jacobian matrix. The nodal displacement matrix $[U_1 \ V_1 \ U_2 \ V_2 \ldots]^T$ transforms to $\{g\}$ by the matrix

$$\begin{bmatrix} N_{1,\xi} & 0 & \cdots \\ N_{1,\eta} & 0 & \cdots \\ 0 & N_{1,\xi} & 0 & \cdots \\ 0 & N_{1,\eta} & 0 & \cdots \end{bmatrix} \qquad \text{(F40.3)}$$

Consequently, the B-matrix is given by the product of the matrices (in the given order) explicitly appearing in the above equations (F40.1), (F40.2) and (F40.3).

Below we present the program which derives the B-matrix for a number of the shape functions, provided these and the nodes are prescribed. The program merely multiplies the above matrices. The arguments of the f-procedure are the number of nodes n and the shape functions $N[i]$, $i = 1, \ldots, n$. The arrays A, S and W are the matrices appearing in (F40.1), (F40.2) and (F40.3), respectively.

We first prepare the array A and the initial form of the array S:

```
> with(linalg,inverse,multiply):
> S:=array(sparse,1..4,1..4):
> A:=array([[1,0,0,0],[0,0,0,1],[0,1,1,0]]):
```

and then state the procedure:

```
> f:=proc()
> m:='m':k:='k':
> x:=sum(N[m]*X[m],m=1..n);y:=sum(N[k]*Y[k],k=1..n);
> J:=array([[diff(x,e),diff(y,e)],[diff(x,h),
    diff(y,h)]]);
```

```
> invJ:=inverse(J):J:='J':
> for i to 2 do for j to 2 do S[i,j]:=invJ[i,j] od:od:
> for i from 3 to 4 do for j from 3 to 4 do S[i,j]:=invJ[i-2,j-2]
    od:od:
> a:=multiply(A,S):A:='A':S:='S':
> W:=array(sparse,1..4,1..n*2):
> ii:=0:for i to n do W[1,i+ii]:=diff(N[i],e):ii:=i od:
> ii:=0:for i to n do W[2,i+ii]:=diff(N[i],h):ii:=i od:
> ii:=1:for i to n do W[3,i+ii]:=diff(N[i],e):ii:=i+1 od:
> ii:=1:for i to n do W[4,i+ii]:=diff(N[i],h):ii:=i+1 od:
> B:=multiply(a,W):a:='a':W:='W':
> end:
```

Consider, for example, the quadrilateral element shown in Fig. F32. Thus, $n = 4$ and the shape functions are given by (F33.3) on the understanding that they are denoted as N[i], $i = 1, 2, 3, 4$, respectively. State the number of nodes and the nodal coordinates (the sequences ee and hh), as follows from Fig. F32:

```
> n:=4:
> ee:=-1,1,1,-1:
> hh:=-1,-1,1,1:
```

and then the shape functions (ξ = e and η = h)

```
> for i to n do e0[i]:=e*ee[i]:h0[i]:=h*hh[i]:
> N[i]:=(1+e0[i])*(1+h0[i])/4 od:
```

It remains to call the above *f*-procedure

```
> f():
```

Now the *B*-matrix may be computed for any values of the nodal coordinates, and, for example, we get for the nodal coordinates employed in Section F34,

```
> nodes:=[X[1]=1,Y[1]=1,X[2]=4,Y[2]=2,X[3]=3,Y[3]=4,
    X[4]=1,Y[4]=4]:
> subs(nodes,op(B)):
```

In particular, the first row is

$$[\frac{2 - 2h}{\%1}, 0, -\frac{-3h + 3}{\%1}, 0, -\frac{3h - e + 2}{\%1}, 0, \frac{-e + 2h + 3}{\%1}, 0]$$

$$\%1 := 3h + 2e - 13$$

A similar approach applies to other elements. File vfem49 deals with an automated derivation of the *B*-matrix in case of a triangular element for a plane–stress problem, while vfem50 and vfem51 with a quadrilateral element and a triangular element, respectively, for a scalar problem.

By specifying the number of nodes n, the shape functions N[i], $i = 1, 2, \ldots, n$, and the nodes themself, one may create the database which contains the B-matrix for a variety of elements.

F41. Concluding remarks

This chapter has dealt with the core of finite element methodology, putting aside a variety of extensions which make this technique even more versatile. At the same time, there is a price to pay: the accuracy evaluation is much more involved than in the case of more traditional methods. Besides its physical and methodological aspects considered in the above, the finite element method has to cope with data and programming. These 'technical' aspects acquire a major importance in the context of this approach. For instance, commercial codes available, written usually in FORTRAN or C, may treat systems with thousands of degrees of freedom. The huge amount of data to process makes necessary special subroutines for mesh generation, solution of very large systems of equations, and computer graphics. Particularly, pre- and postprocessors may occupy a substantial part of the total program.

The following simple example illustrates the generation of a (possibly) non-uniform rectangular or triangular mesh. If z_x and z_y denote the number of divisions in x and y directions, respectively, then the number of elements Z_e is $Z_e = z_x z_y$ and the number of nodes Z_n is $Z_n = (z_x + 1)(z_y + 1)$. Therefore, the arrays Δx_i, $i = 1, 2, \ldots, z_x$ and Δy_j, $j = 1, 2, \ldots, z_y$ provide the global coordinates of each node.

Let the order of numbering be from left to right along the x-axis and from bottom to top along the y-axis. Thus $(0, 0)$ is the origin, $(\Delta x_1, 0)$ is the next node, etc. This construction continues until the arrays Δx_i and Δy_j are exhausted simultaneously.

Since the node numbering affects the matrix bandwidth and the need for renumbering frequently arises, an efficient mesh generator may be highly cost-effective in practical applications. That is why significant efforts have been made in this direction.

Because of its versatility, the finite element method applies to problems dealing directly with modern technology. Particularly, in the framework of computer-aided design this method may considerably improve the quality of engineering decisions.

S. Finite Difference Methods

The basic idea of finite difference techniques stems from the observation that for a finite but sufficiently small increment of the argument (independent variable), for now Δx, the ratio $\Delta f/\Delta x$ (with $f(x)$ being the dependent variable and x the independent one) may deliver a good numerical agreement with the derivative $f_{,x}(x)$. Thus, a differential equation may be replaced by a system of algebraic equations.

Due to the advanced capabilities of computers this simple suggestion, which goes back to the founders of calculus, such as Euler, gave rise to a versatile method of solution of differential equations with a particularly broad field of applications. A huge literature is now available on this method, which constitutes, among others, the main tool of computational fluid dynamics (CFD). In particular, many non-linear problems are amenable to solution by this method. Below we focus on the basis of finite difference methods. For more advanced techniques the reader should consult special courses.

Nevertheless, there should be no uncertainty about the approximate nature of this modeling, so we usually denote the solution to a differential equation by a capital letter and that obtained by the finite difference technique by a lower-case letter. The considerations below assume a constant step-size, but most of the schemes considered apply to a variable step-size too. The combined application of finite difference and finite element methods enables one to consider time-dependent problems (see Chapter F for example).

In general, the presentation refers to results obtained via vfem-files with the extension .ms (MAPLE). To derive the relevant results with the help of MATHEMATICA the user must run the proper vfem-files with the extension .ma.

S1. FD-operators and their accuracy

Let m be an arbitrary integer and $f(x)$ a function of x whose values taken at $x_m = Xm$ are denoted as f_m

194   FINITE DIFFERENCE METHODS

$$f_m = f(x_m) \tag{S1.1}$$

where X is a positive constant. In other words, (S1.1) specifies a discretization of the 'conventional' function $f(x)$, which is assumed to comply with certain conditions of regularity (see Fig. S1).

Now *define* three types of finite differences

$$\Delta_1 f = f_{m+1} - f_m \tag{S1.2a}$$
$$\Delta_2 f = f_m - f_{m-1} \tag{S1.2b}$$
$$\Delta_3 f = f_{m+1/2} - f_{m-1/2} \tag{S1.2c}$$

where $\Delta_1 f$ is referred to as the forward difference, $\Delta_2 f$ as the backward difference and $\Delta_3 f$ the central difference. The $\Delta_1 f$ is also referred to as the *forward Euler algorithm* and $\Delta_2 f$ as the *backward Euler algorithm*. The values of the f-function involved in the definition of the central difference may need to be specified by a separate procedure, for example with the help of interpolation. In agreement with the definition of the first derivative of $f(x)$, any of the expressions (S1.2) may serve as its approximation, if divided by X.

Similarly, the second derivative may be approximated in various ways. For example, using the central difference, set

$$\begin{aligned} f_{,xx} &\approx \Delta_3 f_{,x}/X = [(f_{,x})_{m+1/2} - (f_{,x})_{m-1/2}]/X \\ &= [(\Delta_2 f/X)_{m+1/2} - (\Delta_2 f/X)_{m-1/2}]/X \\ &= (f_{m+1} - 2f_m + f_{m-1})/X^2 \end{aligned} \tag{S1.3}$$

where use has been made of (S1.2b) and (S1.2c).

Of course, (S1.2) and (S1.3) and other similar expressions admit various generalizations, which enables one to involve an arbitrary number of discrete values of $f(x)$ taken with a proper 'weight'. Consequently, a criterion is needed to sort out the finite difference approximations with respect to their accuracy. This is considered below.

A Taylor expansion for $f(x)$ about $x = x_m$ is

$$f(x_m + X) = f_{m+1} = f_m + X f_{,x}(x_m) + X^2/2 f_{,xx}(x_m) + \cdots \tag{S1.4}$$

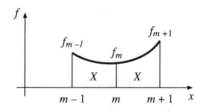

S1 Illustrating discretization of f-function

which gives

$$(f_{m+1} - f_m)/X = \Delta_1 f/X = f_{,x}(x_m) + O(X) \tag{S1.5}$$

Here use has been made of (S1.2a). The error of truncation in the above relation, which describes the inaccuracy introduced by the discrete representation of the first derivative, vanishes as the first power in X. The *order of accuracy* of the approximation associated with (S1.2a) is therefore referred to as the *first* order of accuracy. Similarly, the scheme given by (S1.2b) is also of the first order of accuracy. To evaluate the accuracy of the scheme given by (S1.2c) set two Taylor expansions:

$$f(x_m + X/2) = f_{m+1/2} = f_m + (X/2)f_{,x}(x_m) + (X^2/8)f_{,xx}(x_m) + \cdots \tag{S1.6a}$$
$$f(x_m - X/2) = f_{m-1/2} = f_m - (X/2)f_{,x}(x_m) + (X^2/8)f_{,xx}(x_m) - \cdots \tag{S1.6b}$$

Substracting these relations, we get

$$\Delta_3 f = f_{m+1/2} - f_{m-1/2} = Xf_{,x}(x_m) + O(X^3) \tag{S1.7}$$

and

$$\Delta_3 f/X = f_{,x}(x_m) + O(X^2) \tag{S1.8}$$

This shows that the scheme given by (S1.2c) has the second order of accuracy and from this viewpoint should be given a preference.

The criterion of accuracy just introduced deals with a *local truncation error* as it describes the inaccuracy due to the discrete representation of the derivative at a typical point. Another type of the local error is a *round-off* error which results from the very fact of using a digital computer to perform numerical calculations. The magnitude of the local round-off error depends on the particular hardware used and its adjustment and, unlike the truncation error, is independent of the step-size. The only way to reduce its effect on the accuracy is to make the floating-point representation more precise.

In general, the finite difference method applies to a *grid* over the domain of interest. Depending on the scheme, the local error may evolve in various ways. One therefore needs an approach which would describe a *global* error. This is considered in the next section.

S2. More about accuracy

To illustrate the concept of accuracy in its global sense one should deal with a differential equation and its entire domain of definition rather than with a finite difference operator. For example, consider

$$F_{,x}(x) + bF(x) = 0 \tag{S2.1}$$

196 FINITE DIFFERENCE METHODS

defined for $0 \leq x \leq 1$ and subjected to the normalized initial condition

$$F(x = 0) = 1 \tag{S2.2}$$

Clearly, the exact solution is

$$F(x) = \exp(-bx) \tag{S2.3}$$

Attempting to solve this equation by a finite difference scheme and applying the forward operator Δ_1 (see S1.2a), first get for the elementary step

$$[f(x+X) - f(x)]/X + bf(x) = 0 \tag{S2.4}$$

where X is the step-size. Let M be the total number of intervals, so that $X = 1/M$. Then (S2.4) yields the following recursive relation

$$f_m = f_{m-1}(1 - bX) \tag{S2.5}$$

which after repetitions gives

$$\begin{aligned} f_m &= f_{m-1}(1 - bX) = f_{m-2}(1 - bX)(1 - bX) = \cdots \\ &= f_0(1 - bX)^m \end{aligned} \tag{S2.6}$$

Since, in view of (S2.2), $f_0 = 1$, we get

$$f_m = (1 - bX)^m \tag{S2.7}$$

with

$$m = Xm/X = x_m/X, \quad m = 1, 2, \ldots, M+1 \tag{S2.8}$$

Thus, (S2.3) and (S2.7) represent the exact and approximate solutions, respectively, and the error U at $x = x_m$ is given by their difference

$$U(x_m) = \exp(-bx_m) - (1 - bX)^{(x_m/X)} \tag{S2.9}$$

where use has been made of (S2.8). Recalling that the basic idea requires the step X be a small value, we get a Taylor expansion for the last term of (S2.9) (xm = x_m):

```
> a:= (1 - b* X)^(xm / X);
```

$$a := (1 - b X)^{\left(\frac{xm}{X}\right)}$$

```
> taylor(a,X=0,3);
```

$$\exp(-xm\,b) - 1/2\,\exp(-xm\,b)\,xm\,b^2\,X + O(X^2)$$

Thus, the error at x_m which is given by (S2.9) reduces to

$$U(x_m) = \exp(-bx_m)Xb^2x_m/2 + O(X^2) \tag{S2.10}$$

Normalizing this by the exact solution given by (S2.3), we get

$$U(x_m)/F(x_m) = Xb^2 x_m/2 + O(X^2) = O(X) \qquad \text{(S2.11)}$$

which means that the accuracy in the *global* sense is of the *first* order.

Moreover, (S2.9) or (S2.10) enables one to investigate the error evolution with the increasing number of grid points m. In particular, for $m = 1$ the error is in fact of the local nature. Substituting $x_m = x_1 = X$ into (S2.11), we get

$$U(x_1)/F(x_1) = X^2 b^2/2 + \cdots = O(X^2) \qquad \text{(S2.12)}$$

which indicates the *second* order of accuracy.

Note that these considerations do not account for rounding errors which may play a substantial role in the accuracy of the obtained solution, if the number of calculations is large. Indeed, accumulation of the round-off error follows the simple rule:

$$e^r_{\text{global}} = M e^r_{\text{local}} = e^r_{\text{local}}/X \qquad \text{(S2.13)}$$

where e^r is the round-off error. This shows that, unlike the local round-off error, the global one does depend on the step-size.

S3. Sample problem

(MAPLE file vfem53)

Consider a model equation

$$F_{,x}(x) = -6F(x) + 5\,\mathrm{e}^{-x} = g(F, x) \qquad \text{(S3.1)}$$

subjected to the initial condition

$$F(x = 0) = 1 \qquad \text{(S3.2)}$$

Applying the forward Euler algorithm, replace (S3.1) by

$$f_{m+1} = f_m + X[-6f_m + 5\exp(-x_m)] \qquad \text{(S3.3)}$$

Similarly, the backward Euler algorithm yields

$$f_{m+1} = f_m + X[-6f_{m+1} + 5\exp(-x_{m+1})] \qquad \text{(S3.4)}$$

It is seen that (S3.3) assumes $g(F, x)$ to be 'locally constant' at the $(m+1)$th step with value f_m, while (S3.4) assumes this quantity to be 'locally constant' at the $(m+1)$th step with value f_{m+1}.

The forward Euler algorithm is obviously an *explicit* computational scheme, while the backward Euler algorithm is an *implicit* scheme, as f_{m+1} appears on both sides of the equation. Generally speaking, implicit schemes may provide a more accurate solution, though they are more difficult to compute.

Let us compute the forward scheme (S3.3) for the interval $0 \le x \le 3$. First, specify the step size, say, $X = 0.3$, and then compute (S3.3) with the help of a do-command:

198 FINITE DIFFERENCE METHODS

```
> f[0]:=1:X:=0.3:x[0]:=0:
> for m to 10 do x[m]:=x[0]+m*X:
> f[m]:=f[m-1]+X*(-6*f[m-1] +5*exp(-x[m-1])):od:
```

and plot the results along with the exact solution $F(x) = \exp(-x)$

```
> plot({[[x[i],f[i]]$ i=0..10],exp(-x)},x=0..3);
```

(see Fig. S2).

Adopting a smaller step-size, $X = 0.1$, one gets more accurate results:

```
> X:=0.1:
> for m to 30 do x[m]:=x[0]+m*X:
> f[m]:=f[m-1]+X*(-6*f[m-1] +5*exp(-x[m-1])):od:
```

Plotting these results is left to the reader.

Now consider the backward Euler algorithm given by (S3.4). In this case the unknown value appears on both sides of the equation. Consequently, one must resolve a system of simultaneous equations, the operation absent in the previous forward scheme. Specifying the step size as $X = 0.3$, we get

```
> f:='f':m:='m':
> f[0]:=1:X:=0.3:x[0]:=0:
> for m to 10 do x[m]:=x[0]+m*X:
> Q[m]:=-f[m]+f[m-1]+X*(-6*f[m] +5*exp(-x[m])):
  od:Digits:=5:
> solve({Q[n]$ n=1..10},{f[n]$ n=1..10});
```

{f[1] = .75399, f[2] = .56328, f[3] = .41896, f[4] = .31098,
   f[5] = .23060, f[6] = .17091, f[7] = .12664, f[8] = .093828,
      f[9] = .069514, f[10] = .051500}

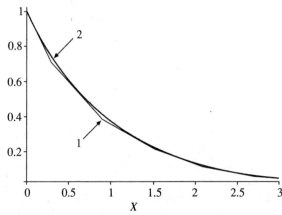

S2 Approximate (1) and exact (2) solutions

Once again, plotting the results is left to the reader as an exercise.

These results show that this implicit scheme provides a better accuracy than the explicit forward algorithm. In general, because of the presence of unknowns on both sides of the equations which may be interpreted as a kind of feedback, the implicit schemes may deliver an improved accuracy at the expense of more involved computational procedure.

A natural generalization of the above schemes is the so-called *trapezoidal* scheme. Consider the equation of the type given by (S3.1). The trapezoidal scheme assumes the field function $g(F, x)$ be locally equal to the average of its adjacent values:

$$f_{m+1} = f_m + (X/2)[g(f_m, x_m) + g(f_{m+1}, x_{m+1})] \tag{S3.5}$$

Properly editing one of the above files, one may obtain the solution to (S3.1) with the help of this algorithm too.

S4. Runge–Kutta methods, 1

Consider the first-order equation in its fairly general form

$$Y_{,x}(x) = g(x, Y) \tag{S4.1}$$

and in the spirit of the above relations set up the basic relation of the so-called Runge–Kutta method as

$$y_{m+1} = y_m + X\phi(x_m, y_m, X) \tag{S4.2}$$

where y_m is assumed to approximate $Y(x_m)$ and $\phi(x_m, y_m, X)$ to approximate $g(x, Y)$ as $x_m \leq x \leq (x_m + X)$. The expression (S4.2) may be viewed as a generalization of the Euler method. The derivation of $\phi(x_m, y_m, X)$ requires a considerable amount of algebra and it would be useful to set a convenient system of notation.

Let k_1 and k_2 be defined by

$$\begin{aligned} k_1 &= g(x_m, y_m) \\ k_2 &= g(x_m + pX, y_m + qXk_1) = g(x_m + pX, y_m + qXg(x_m, y_m)) \end{aligned} \tag{S4.3}$$

with p and q being unspecified yet constants. Further, assume the above ϕ-function to be a weighted average of these quantities by

$$\phi(x_m, y_m, X) = ak_1 + bk_2 \tag{S4.4}$$

with the as yet unspecified 'weights' a and b.

Expanding k_2 in a Taylor's series as a function of its two variables, we get

$$\begin{aligned} k_2 &= g(x_m + pX, y_m + qXk_1) = g(x_m, y_m) + pXg_{,x}(x_m, y_m) \\ &\quad + qXg(x_m, y_m)g_{,y}(x_m, y_m) + O(X^2) \end{aligned} \tag{S4.5}$$

Substituting the expressions for k_1 and k_2 in (S4.4) and then in (S4.2), we get

$$y_{m+1} = y_m + X\phi(x_m, y_m, X) = y_m + X(ak_1 + bk_2)$$
$$= y_m + X[ag(x_m, y_m) + bg(x_m, y_m) + bpXg_{,x}(x_m, y_m)$$
$$+ bqXg(x_m, y_m)g_{,y}(x_m, y_m)] + \cdots \quad \text{(S4.6)}$$

or, in a form of series with respect to the step-size X,

$$y_{m+1} = y_m + X[ag(x_m, y_m) + bg(x_m, y_m)]$$
$$+ X^2[bpg_{,x}(x_m, y_m) + bqg(x_m, y_m)g_{,y}(x_m, y_m)] + O(X^3)$$

It remains to identify the coefficients a, b, p and q. To this end, represent $Y(x)$ as a Taylor's series about $x = x_m$:

$$Y(x_m + X) = Y(x_m) + Xg(x_m, Y_m) + (X^2/2)g'(x_m, Y_m) + O(X^3) \quad \text{(S4.8)}$$

where by the chain rule

$$g'(x_m, Y_m) = g_{,x}(x_m, Y_m) + g_{,y}(x_m, Y_m)g(x_m, Y_m) \quad \text{(S4.9)}$$

Substituting (S4.9) into (S4.8) gives

$$Y(x_m + X) = Y(x_m) + Xg(x_m, Y_m) + (X^2/2)[g_{,x}(x_m, Y_m)$$
$$+ g_{,y}(x_m, Y_m)g(x_m, Y_m)] + O(X^3) \quad \text{(S4.10)}$$

Comparing this expression with (S4.7) and setting $y_m = Y_m = Y(x_m)$, one observes that it is possible to equate the terms in like powers of X. This gives the following two equations:

$$ag(x_m, y_m) + bg(x_m, y_m) = g(x_m, y_m)$$
$$bpg_{,x}(x_m, y_m) + bqg(x_m, y_m)g_{,y}(x_m, y_m)$$
$$= g_{,x}(x_m, Y_m)/2 + g_{,y}(x_m, Y_m)g(x_m, Y_m)/2 \quad \text{(S4.11)}$$

In turn, the first of these equations yields $a + b = 1$ and the second yields $bp = 1/2$, $bq = 1/2$. Thus, one of the coefficients, say b, remains an arbitrary value. The conventional choice is $b = 1$ or $b = 1/2$ and two possible sets of the coefficients are

$$\begin{aligned} b &= 1, & a &= 0, & p &= q = 1/2 \\ b &= a = 1/2, & & & p &= q = 1 \end{aligned} \quad \text{(S4.12)}$$

Substituting this into (S4.6), one gets the following versions of the *second order* Runge–Kutta algorithm:

(1) $b = 1$

$$y_{m+1} = y_m + X[g(x_m, y_m) + (X/2)g_{,x}(x_m, y_m)$$
$$+ (X/2)g(x_m, y_m)g_{,y}(x_m, y_m)] + \cdots$$
$$= y_m + Xg(x_m + X/2, y_m + Xg(x_m, y_m)/2) \quad \text{(S4.13)}$$

(2) $b = 1/2$

$$y_{m+1} = y_m + (X/2)[g(x_m, y_m) + g(x_m, y_m) + Xg_{,x}(x_m, y_m) \\ + Xg(x_m, y_m)g_{,y}(x_m, y_m)] + \ldots \\ = y_m + (X/2)[g(x_m, y_m) + g(x_m + X, y_m + Xg(x_m, y_m))] \quad (S4.14)$$

In a similar way one may construct the higher order Runge–Kutta schemes, this similarity being a substantial advantage of the method. This is considered in the next section.

S5. Runge–Kutta methods, 2

(*MAPLE file vfem54*)

In case of the third order algorithm, the basic relation is set up as

$$y_{m+1} = y_m + X\phi(x_m, y_m, X) \quad (S5.1)$$

where

$$\phi(x_m, y_m, X) = ak_1 + bk_2 + ck_3 \quad (S5.2)$$

and

$$k_1 = g(x_m, y_m) \\ k_2 = g(x_m + pX, y_m + pXk_1) \quad (S5.3) \\ k_3 = g[x_m + rX, y_m + sXk_1 + (r-s)Xk_2]$$

The six coefficients a, b, c, p, r and s follow, as earlier, from comparison of (S5.1) with a Taylor's expansion for $Y(x)$. One arrives at the following four equations:

$$a + b + c = 1, \quad bp + cr = 1/2 \\ bp^2 + cr^2 = 1/3, \quad cps = 1/6 \quad (S5.4)$$

which leaves two of the constants arbitrary and gives rise to a variety of schemes. For a particular scheme suggested by Kutta the final results are

$$y_{m+1} = y_m + X(k_1 + 4k_2 + k_3)/6 \quad (S5.5)$$

where

$$k_1 = g(x_m, y_m) \\ k_2 = g(x_m + X/2, y_m + k_1 X/2) \quad (S5.6) \\ k_3 = g[x_m + X, y_m - Xk_1 + 2Xk_2]$$

202 FINITE DIFFERENCE METHODS

In case of the most common fourth-order algorithm a widely used scheme is given by

$$y_{m+1} = y_m + X(k_1 + 2k_2 + 2k_3 + k_4)/6 \tag{S5.7}$$

where

$$\begin{aligned} k_1 &= g(x_m, y_m) \\ k_2 &= g(x_m + X/2, y_m + Xk_1/2) \\ k_3 &= g(x_m + X/2, y_m + Xk_2/2) \\ k_4 &= g(x_m + X, y_m + Xk_3) \end{aligned} \tag{S5.8}$$

Since the nth order Runge–Kutta method is derived by the matching procedure with the Taylor's series for the solution $Y(x)$ through terms of order X^n, the local truncation error R admits the following representation:

$$R = AX^{n+1} + O(X^{n+2}) \tag{S5.9}$$

where constant A depends on the properties of $g(x, Y)$. Note that the evaluation of this constant is not an easy thing to do. One of the empiric ways is as follows.

Integrate between two points x_m and x_{m+1} using two different steps X_1 and X_2 and denote the solutions as $y_{m+1}^{(1)}$ and $y_{m+1}^{(2)}$, respectively. Assuming that R makes the dominant contribution in the total error for this step and that the 'exact' solution is Y_{m+1}, set up the estimates

$$\begin{aligned} Y_{m+1} - y_{m+1}^{(1)} &= AX_1^{n+1}(x_{m+1} - x_m)/X_1 \\ Y_{m+1} - y_{m+2}^{(1)} &= AX_2^{n+1}(x_{m+1} - x_m)/X_2 \end{aligned} \tag{S5.10}$$

which may be solved for both A and Y_{m+1}. It should be noted that this procedure, if frequently applied, may greatly increase the number of calculations and computing time.

The family of Runge–Kutta schemes can be shown to provide results convergent to the exact solution with $X \to 0$. Nevertheless, the choice of X should be made on the basis of compromise in order to keep rounding errors, which increase with the number of operations performed, under control.

Now consider a system of n simultaneous first order ordinary differential equations:

$$\begin{aligned} Y_{1,x}(x) &= g_1(x, Y_1, Y_2, \ldots, Y_n) \\ Y_{2,x}(x) &= g_2(x, Y_1, Y_2, \ldots, Y_n) \\ &\cdots\cdots\cdots\cdots\cdots\cdots\cdots\cdots \\ Y_{n,x}(x) &= g_n(x, Y_1, Y_2, \ldots, Y_n) \end{aligned} \tag{S5.11}$$

subjected to the initial conditions at $x = x_0$

$$Y_1(x_0) = Y_{1,0}; \quad Y_2(x_0) = Y_{2,0}; \quad \ldots; \quad Y_n(x_0) = Y_{n,0} \tag{S5.12}$$

Note that a higher-order differential equation admits representation in the form given by (S5.11) with properly reformulated initial conditions. The Runge–Kutta method treats the above system too with the following difference only: the method applies in parallel to each of the equations at each of the steps.

As a first application of the Runge–Kutta method, consider again the simple equation (S3.1) reproduced below:

$$F_{,x}(x) = -6F(x) + 5e^{-x} \qquad (S3.1)$$

which is subjected to the initial condition $F(x=0) = 1$.

Following (S5.7) and (S5.8), state the right-hand side of (S3.1), the initial condition, the step-size $X = 0.3$ and then with the help of do-command the Runge–Kutta algorithm itself

```
> g(xx,yy):=-6*yy+5*exp(-xx);
> y[0]:=1:x[0]:=0:X:=0.3:
> for m to 10 do
> x[m]:=x[0]+m*X:
> k1:=subs([xx=x[m-1],yy=y[m-1]],g(xx,yy)):
> k2:=subs([xx=x[m-1]+X/2,yy=y[m-1]+X*k1/2],g(xx,yy)):
> k3:=subs([xx=x[m-1]+X/2,yy=y[m-1]+X*k2/2],g(xx,yy)):
> k4:=subs([xx=x[m-1]+X,yy=y[m-1]+X*k3],g(xx,yy)):
> y[m]:=y[m-1]+X*(k1+2*k2+2*k3+k4)/6:od:
```

The results, which are of a good accuracy, may be conviniently represented with the help of plot and compared with the previous solutions, which is left to the reader.

S6. Transient vibrations

(MAPLE file vfem55)

For the system of n simultaneous equations (see (S5.11)) the Runge–Kutta algorithm given by (S5.7) and (S5.8) takes the form

$$y^j_{m+1} = y^j_m + X(k^j_1 + 2k^j_2 + 2k^j_3 + k^j_4)/6 \qquad (S6.1)$$

where $j = 1, 2, \ldots, n$ and

$$\begin{aligned}
k^j_1 &= g^j(x_m, y^1_m, y^2_m, \ldots, y^n_m) \\
k^j_2 &= g^j(x_m + X/2, y^1_m + Xk^1_1/2, \ldots, y^n_m + Xk^n_1/2) \\
k^j_3 &= g^j(x_m + X/2, y^1_m + Xk^1_2/2, \ldots, y^n_m + Xk^n_2/2) \\
k^j_4 &= g^j(x_m + X, y^1_m + Xk^1_3, \ldots, y^n_m + Xk^n_3)
\end{aligned} \qquad (S6.2)$$

204 FINITE DIFFERENCE METHODS

As an example, consider a transient behaviour of the classical oscillator

$$MZ_{,xx}(x) + CZ_{,x}(x) + KZ(x) = 0 \tag{S6.3}$$

with the initial conditions given by

$$Z(x=0) = z_0, \quad Z_{,x}(x=0) = 0 \tag{S6.4}$$

This equation of the second order admits the formulation in a form of two simultaneous equations of the first order (see S5.11). Denote

$$Z(x) = Y_1, \quad Z_{1,x} = Y_2 \tag{S6.5}$$

and state (S6.3) as

$$Y_{2,x} = -(C/M)Y_2(x) - (K/M)Y_1(x)Y_{1,x} = Y_2 \tag{S6.6}$$

For illustration purposes, assume that the constants M, C, K and z_0 are all equal to 1 (in a dimensionless reference system) and the interval for x is $(0, 4)$. Anticipating a rapidly varying function, specify the step-size as $X = 0.005$ and then program the Runge–Kutta algorithm given by (S6.1) and (S6.2). Besides the constants, the program below may apply to an arbitrary system of two simultaneous equations

```
> M:=1:C:=1:K:=1:
> X:=0.005:x[0]:=0:
> y1[0]:=1:y2[0]:=0:
> g1(xx,yy1,yy2):=yy2:
> g2(xx,yy1,yy2):=-(C/M)*yy2-(K/M)*yy1:
> m:='m':
> for m to 800 do
> x[m]:=x[0]+m*X:
> k11:=subs([xx=x[m-1],yy1=y1[m-1],
   yy2=y2[m-1]],g1(xx,yy1,yy2)):
> k12:=subs([xx=x[m-1],yy1=y1[m-1],
   yy2=y2[m-1]],g2(xx,yy1,yy2)):
> k21:=subs([xx=x[m-1]+X/2,yy1=y1[m-1]+X*k11/2,
   yy2=y2[m-1]+X*k12/2],g1(xx,yy1,yy2)):
> k22:=subs([xx=x[m-1]+X/2,yy1=y1[m-1]+X*k11/2,
   yy2=y2[m-1]+X*k12/2],g2(xx,yy1,yy2)):
> k31:=subs([xx=x[m-1]+X/2,yy1=y1[m-1]+X*k21/2,
   yy2=y2[m-1]+X*k22/2],g1(xx,yy1,yy2)):
> k32:=subs([xx=x[m-1]+X/2,yy1=y1[m-1]+X*k21/2,
   yy2=y2[m-1]+X*k22/2],g2(xx,yy1,yy2)):
> k41:=subs([xx=x[m-1]+X,yy1=y1[m-1]+X*k31,
   yy2=y2[m-1]+X*k32],g1(xx,yy1,yy2)):
> k42:=subs([xx=x[m-1]+X,yy1=y1[m-1]+X*k31,
   yy2=y2[m-1]+X*k32],g2(xx,yy1,yy2)):
```

```
> y1[m]:=y1[m-1]+X*(k11+2*k21+2*k31+k41)/6:
> y2[m]:=y2[m-1]+X*(k12+2*k22+2*k32+k42)/6:
> od:
```

Equation (S6.3) with the conditions (S6.4) can be easily resolved in a conventional way as follows:

```
> eq:=diff(z(x),x,x)+diff(z(x),x)+z(x)=0;
> dsolve({eq,z(0)=1,D(z)(0)=0},z(x));
```

$$z(x) = \exp(-1/2\,x)\cos(1/2\,3^{1/2}x) + 1/3\,3^{1/2}\exp(-1/2\,x)\sin(1/2\,3^{1/2}x)$$

which provides an opportunity to evaluate the accuracy of the Runge–Kutta solution. The plot in Fig. S3 illustrates a very good agreement as the two solutions almost coincide:

```
> z(x):=op(2,"):
> plot({[[x[i],y1[i]]$ i=0..800],z(x)},x=0..4);
```

To appreciate this high accuracy, note that the 'period' of vibrations is of the order 2π, which is large compared to the step-size $X = 0.005$. The reader may increase X, repeat the calculations and then observe the decay in the accuracy of results. The disparity between the derivatives of the approximate and exact solutions becomes even greater.

S7. Multi-step schemes

The schemes investigated so far involve two values only of the unknown function at each of the steps. Some of these so-called *single-step* algorithms, for example, the Runge–Kutta method, may be extremely accurate if the step-size is sufficiently small. Nevertheless, this family of methods may not always be

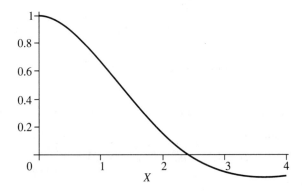

S3 Approximate and exact solutions to vibration problem

computationally efficient. In order to illustrate this point, note that the fourth-order Runge–Kutta method involves four evaluations of the g-function per step.

An alternative, so-called n-step, approach involves n values of the unknown function, $y_i, y_{i-1}, \ldots, y_{i-n+1}$ to evaluate y_{i+1} and has the following generic form:

$$y_{i+1} = a_0 y_i + a_1 y_{i-1} + \cdots + a_{n-1} y_{i-n+1} + X[b_{-1} g(y_{i+1}, x_{i+1})$$
$$+ b_0 g(y_i, x_i) + b_1 g(y_{i-1}, x_{i-1}) + \cdots + b_{n-1} g(y_{i-n+1}, x_{i-n+1})] \quad (S7.1)$$

where x_{i-k} are evenly spaced with the step X and the a- and b-coefficients need to be specified by a separate procedure. The form (S7.1) follows from interpolation or quadrature rules, both having much to do with the subject of approximate solution to differential equations. The a-coefficients in (S7.1) should take into account the 'history' of $Y(x)$, while the b-coefficients the effect of g-function. Note, this *multi-step* algorithm is implicit unless $b_{-1} = 0$. Though multi-step methods may be very efficient, this is no certainty, and their comparison with single-step methods depends on a particular problem.

There are together $(2n+1)$ coefficients in (S7.1). A general approach of specifying the coefficients is to require the integration scheme to be exact for a polynomial of order N or less. This seems to be a natural criterion, as the polynomials represent a broad family of functions.

The *Adams–Bashforth scheme* first sets in (S7.1) $b_{-1} = 0$, $n = N$ and $a_k = 0$, $k = 1, \ldots, N-1$. Then (S7.1) acquires the form

$$y_{i+1} = a_0 y_i + X[b_0 g(y_i, x_i) + b_1 g(y_{i-1}, x_{i-1}) + \cdots$$
$$+ b_{N-1} g(y_{i-N+1}, x_{i-N+1})] \quad (S7.2)$$

The remaining coefficients are specified so as to make the integration scheme (S7.2) exact for a polynomial of order N or less.

Let us derive the *Adams–Bashforth scheme* of the first order, setting $N = 1$. Then (S7.2) becomes

$$y_{i+1} = a_0 y_i + X b_0 g(y_i, x_i) \quad (S7.3)$$

Assuming the polynomial $Y = c_0 + c_1 x$, and $g = Y_{,x} = c_1$ and substituting in (S7.3), we get

$$c_0 + c_1 x_{i+1} = a_0 c_0 + a_0 c_1 x_i + X b_0 c_1 \quad (S7.4)$$

which gives

$$c_0(1 - a_0) = 0 \quad \text{and} \quad c_1(x_{i+1} - a_0 x_i - X b_0) = 0 \quad (S7.5)$$

Thus, the coefficients are

$$a_0 = 1, \quad b_0 = (x_{i+1} - x_i)/X = 1 \quad (S7.6)$$

and (S7.3) becomes

$$y_{i+1} = y_i + X g(y_i, x_i) \quad (S7.7)$$

which recovers the forward Euler algorithm. One may derive in a similar way the Adams–Bashforth scheme of the Nth order.

The *Adams–Moulton scheme* assumes $n = N - 1$ and $a_k = 0$, $k = 1, \ldots, N - 1$. Then (S7.1) becomes

$$y_{i+1} = a_0 y_i + X[b_{-1} g(y_{i+1}, x_{i+1})$$
$$+ b_0 g(y_i, x_i) + b_1 g(y_{i-1}, x_{i-1}) + \cdots + b_{N-2} g(y_{i-N+2}, x_{i-N+2})] \quad (S7.8)$$

where once again the remaining coefficients are specified so as to make the integration scheme (S7.8) exact for a polynomial of order N or less.

Finally, for the *Gear scheme* set $n = N$ and $b_k = 0$, $k = 1, \cdots, N$, which yields

$$y_{i+1} = a_0 y_i + a_1 y_{i-1} + \ldots + a_{N-1} y_{i-N+1} + X b_{-1} g(y_{i+1}, x_{i+1}) \quad (S7.9)$$

with the same prescription for specifying the remaining coefficients.

Note that the multi-step algorithms need, unlike the Runge–Kutta method, a special start-up procedure, as they include the 'previous history' of the function which may not be given by the initial conditions. For example, the fourth-order Adams–Bashforth algorithm involves the relevant quantities at the five points, from $(i - 3)$ up to $(i + 1)$. This and other explicit algorithms are given in the next section which deals with the automatic derivation of multi-step schemes.

S8. Automatic generation of multi-step schemes

(*MAPLE files vfem56, 57 and 58*)

The derivation of particular multi-step algorithms is tedious work and it is natural to apply a computer algebra code for this purpose.

Below we present programs which automatically deduce the explicit formulae of various orders. The program consist of two parts. The first one formulates a polynomial of the order N and its first derivative, which are denoted as yy and gg, respectively, and then states the basic expression (S7.1). For sake of convenience, a reference point i in (S7.1) is taken below as $i = 1$. The second part states the specific values of coefficients typical of the particular scheme, then specifies the rest of them by solving the system of equations to comply with the 'polynomial criterion' and finally states the algorithm.

We begin with the *Adams–Bashforth algorithm*, formulating the first part with the help of proc to ensure automatic derivation

```
> A_B:=proc(N)
> yy:=sum(c[kk]*x^kk,kk=0..N):
> gg:=diff(yy,x):
> for j from (2-N) to 2 do y[j]:=subs(x=x[1]+X*(j-1),yy):
> g[j]:=subs(x=x[1]+X*(j-1),gg) od:
> s:=expand(y[2]-sum(a[i]*y[-i+1],i=0..N-1)
    -X*(sum(b[ii]*g[-ii+1],ii=-1..N-1))):
```

208   FINITE DIFFERENCE METHODS

where $g[j] = g(y_j, x_j)$.

The second part of this program states the conditions of the Adams–Bashforth scheme with regard to the coefficients of (S7.1), as explained in Section S7:

```
> b[-1]:=0:
> for z to N do a[z]:=0 od:
```

and specifies the remaining coefficients of (S7.1), which are then substituted back in the generic expression (S7.1) to produce the final form of the algorithm:

```
> for p from 0 to N do eq[p]:=coeff(s,c[p]) od:
> S:=solve({eq[r]$ r=0..N},{a[0],b[t]$ t=0..N-1});
> y:='y':g:='g':
> y[k+1]=sum(a[i]*y[k-i],i=0..N-1)+X*(sum(b[i]*g[k-i],
    i=-1..N-1)):
> subs(S,");
> end;
```

All that is left to do is to call the procedure A_B(N) where N indicates the order of the algorithm. For example, we get the first-order algorithm:

```
> A_B(1);
```

$$y[k+1] = y[k] + X\, g[k]$$

which coincides with (S7.7). Now we get the fourth-order algorithm

```
> A_B(4);
```

$$y[k+1] = y[k] + X\left(\frac{55}{24}g[k] - \frac{59}{24}g[k-1] + \frac{37}{24}g[k-2] - 3/8g[k-3]\right)$$

Similarly, one constructs the programs for the Gear and Adams–Moulton schemes, which are given below as the G-procedure and A_M-procedure, respectively.

For the *Gear algorithm*:

```
G:=proc(N)
yy:=sum(c[kk]*x^kk,kk=0..N):
gg:=diff(yy,x):
for j from (2-N) to 2 do y[j]:=subs(x=x[1]+X*(j-1),yy):
g[j]:=subs(x=x[1]+X*(j-1),gg) od:
s:=expand(y[2]-sum(a[i]*y[-i+1],i=0..N-1)
    -X*(sum(b[ii]*g[-ii+1],ii=-1..N-1))):
for z from 0 to N-1 do b[z]:=0 od:
for p from 0 to N do eq[p]:=coeff(s,c[p]) od:
S:=solve({eq[r]$ r=0..N},{b[-1],a[t]$ t=0..N-1});
```

```
y:='y':g:='g':
y[k+1]=sum(a[i]*y[k-i],i=0..N-1)+X*(sum(b[i]*g[k-i],
   i=-1..N-1)):
subs(S,'');
end;
```

For example, deduce Gear's scheme of the third order by

```
> G(3);
```

$$y[k+1] = \frac{18}{11}y[k] - 9/11y[k-1] + 2/11y[k-2] + 6/11Xg[k+1]$$

For the *Adams–Moulton algorithm*:

```
A_M:=proc(N)
K:=N+1:
yy:=sum(c[kk]*x^kk,kk=0..K):
gg:=diff(yy,x):
for j from (2-N) to 2 do y[j]:=subs(x=x[1]+X*(j-1),yy):
   g[j]:=subs(x=x[1]+X*(j-1),gg) od:
s:=expand(y[2]-sum(a[i]*y[-i+1],i=0..N-1)
   -X*(sum(b[ii]*g[-ii+1],ii=-1..N-1))):
for z to N do a[z]:=0 od:
for p from 0 to K do eq[p]:=coeff(s,c[p]) od:
S:=solve({eq[r]$ r=0..K},{a[0],b[t]$ t=-1..N-1});
y:='y':g:='g':
y[k+1]=sum(a[i]*y[k-i],i=0..N-1)+X*(sum(b[i]*g[k-i],
   i=-1..N-1)):
subs(S,'');
end;
```

Note that for this scheme to derive the results of the *K*th-order one should call on A_M(K-1). For example, get the third order scheme as follows:

```
> s2:=A_M(2);
```

$$s2 := y[k+1] = y[k] + X(5/12\,g[k+1] + 2/3\,g[k] - 1/12\,g[k-1])$$

S9. Start-up for multi-step schemes

(MAPLE file vfem58)

As noted in Section S7, multi-step schemes are not self-starting. Figuratively speaking, they have a certain 'inertia' and need a special start-up procedure as some of the relevant values are not initially known. For example, try to apply the above Adams–Moulton scheme to the model equation treated in Section S3:

FINITE DIFFERENCE METHODS

$$F_{,x}(x) = -6F(x) + 5e^{-x} = g(F, x) \tag{S3.1}$$

subjected to the initial condition

$$F(x = 0) = 1 \tag{S3.2}$$

It is seen that the initial information is not sufficient for calculating the expression s2 given at the end of Section S8, as it contains three unknown values. One of the possible start-up techniques is the use of the scheme of a lower order for the first steps or use of interpolation routines.

In the case at hand, following the first of these possibilities, state first the second-order algorithm by

```
> s1:=A_M(1):

    s1:= y[k + 1] = y[k] + X (1/2 g[k + 1] + 1/2 g[k])
```

assume the step-size $X = 0.3$ and $k = 0$, and use the above (S3.1) and (S3.2) to calculate the unknown function for $k = 1$ from s1. Proceed as follows:

```
> Digits:=5:
> X:=0.3:x[0]:=0:y[0]:=1:
> for k from 0 to 1 do x[k]:=x[0]+X*k:g[k]:=-6*y[k]
  +5*exp(-x[k]):od:
```

and resolve s1 for $y[1]$

```
> k:=0:
> y[1]:=solve(s1,y[1]);

    y[1]:= .73979
```

Now one has at the disposal all the values needed to employ the third-order Adams–Moulton algorithm, as shown below:

```
> k:='k':
> s2:=A_M(2);

    s2:= y[k + 1] = 1. y[k]
         + .12500 g[k + 1] + .20000 g[k] - .025000 g[k - 1]

> for k from 0 to 10 do x[0]:=0;y[0]:=1:y[1]:=0.73979
> x[k]:=x[0]+X*k:g[k]:=-6*y[k]+5*exp(-x[k]):od:
> for k to 9 do Q[k]:=s2 od:
```

It remains to resolve the system of nine simultaneous equations

```
> solve({Q[n]$ n=1..9},{y[m]$ m=2..10});

    {y[2] = .54906, y[3] = .40657, y[4] = .30130, y[5] = .22318,
     y[6] = .16535, y[7] = .12249, y[8] = .090743, y[9] = .067217,
     y[10] = .049804}
```

Note that the multi-step schemes apply to a system of equations in the same way as the Runge–Kutta scheme does. In this case a start-up routine is needed by each of the equations.

S10. Iterative predictor–corrector methods

Differential equations of the second order, which are of particular significance for engineering, may be solved by one of the above methods, if transformed to the system of equations of the first order (see Section S6). As an alternative, we consider the so-called, iterative predictor–corrector method on the example

$$Y_{,xx} = g(Y, Y_{,x}, x) \qquad (S10.1)$$

Among various versions of this method, we present below the *average-acceleration approach*. Begin with the trapezoidal rule given by (S3.5). Adjusting the notations and this time denoting the step-number by superscript, we get for the first derivative

$$y^{m+1} = y^m + (X/2)(y^m_{,x} + y^{m+1}_{,x}) \qquad (S10.2)$$

and for the second derivative

$$y^{m+1}_{,x} = y^m_{,x} + (X/2)(y^m_{,xx} + y^{m+1}_{,xx}) \qquad (S10.3)$$

Together with (S10.1) written in a discrete form

$$y^{m+1}_{,xx} = g(y^{m+1}, y^{m+1}_{,x}, x^{m+1}) \qquad (S10.4)$$

(S10.2) and (S10.3) provide a basis for an approximate solution to (S10.1), if iterations are performed within each step m. To this end, we denote the iteration number as i, and rewrite these equations in a form which explicitly indicates the iterative terms with the help of brackets:

$$[y^{m+1}_{,x}]_i = y^m_{,x} + (X/2)y^m_{,xx} + (X/2)[y^{m+1}_{,xx}]_{i-1} \qquad (S10.5)$$

$$[y^{m+1}]_i = y^m + (X/2)y^m_{,x} + (X/2)[y^{m+1}_{,x}]_i \qquad (S10.6)$$

$$[y^{m+1}_{,xx}]_i = [g(y^{m+1}, y^{m+1}_{,x}, x^{m+1})]_i \qquad (S10.7)$$

and supplement these by Euler's scheme as a start-up procedure

$$y^{m+1}_{,x} = y^m_{,x} + y^m_{,xx} X \qquad (S10.8)$$

To begin, set $m = 0$ and find $y^1_{,x}$ from (S10.8). Note that $y^0_{,x}$ and $y^0_{,xx}$ follow from the initial conditions and (S10.1). Equations (S10.6) and (S10.7) provide the values of $[y^1]_1$ and $[y^1_{,xx}]_1$. Repeat this routine for the next iterations while using (S10.5) instead of (S10.8). In a similar way one can compute the results for the second step, etc. Note that Euler's scheme is employed as a start-up procedure at the beginning of each step.

Consider the transient behaviour of an oscillator described by the equation

$$MY_{,xx}(x) + CY_{,x}(x) + KY(x) = P \quad \text{(S10.9)}$$

with the initial conditions given by

$$Y(x=0) = 0, \quad Y_{,x}(x=0) = 0 \quad \text{(S10.10)}$$

Recasting (S10.9) in the form of (S10.1), we get

$$Y_{,xx} = P/M - (C/M)Y_{,x} - (K/M)Y \quad \text{(S10.11)}$$

Equations (S10.10) and (S10.11) provide

$$y^0_{,xx} = P/M \quad \text{(S10.12)}$$

and (S10.8) and (S10.10) provide

$$y^1_{,x} = PX/M \quad \text{(S10.13)}$$

Now y^1 follows from (S10.6) as

$$[y^1]_1 = PX^2/(2M) \quad \text{(S10.14)}$$

and $y^1_{,xx}$ from (S10.7) and (S10.11) as

$$[y^1_{,xx}]_1 = P/M - PCX/M^2 - PKX^2/(2M^2) \quad \text{(S10.15)}$$

For the second iteration within this step use (S10.5) instead of Euler's rule (S10.8) and repeat the routine, etc. The iterations stop when the desired accuracy is achieved and the procedure turns to the next step.

Equation (S10.8) is an example of a *predictor*, as it explicitly specifies the 'future' in terms of the 'past'. On the other hand, (S10.5) illustrates an implicit *corrector*, as it improves the value of the first derivative, making use of the known value of the second derivative.

The next section deals with applications of this method to non-linear vibrations.

S11. Non-linear vibrations

(MAPLE files vfem62 and 63)

Consider the equation of free non-linear vibrations given by

$$Y_{,xx} + \omega^2 \sin Y = 0 \quad \text{(S11.1)}$$

where ω depends on the system parameters. In the case of a pendulum, for example, with the usual notation, $\omega^2 = g/L$, L being the length, $x = t$, t being time, and $Y = \phi$, ϕ being the angle of rotation. Equation (S11.1) can be resolved exactly in terms of elliptic integrals. This solution shows that for $\omega^2 = 3.43771/s^2$ and the initial condition $Y(x=0) = \pi/2$, $Y_{,x}(x=0) = 0$, the period T equals 4 s, which may be of help in estimating the accuracy of approximate solutions. Therefore we consider below the equation

$$Y_{,xx} = -3.4377 \sin Y \tag{S11.2}$$

with the initial conditions

$$Y(x=0) = \pi/2, \qquad Y_{,x}(x=0) = 0 \tag{S11.3}$$

As Section S10 shows, a program based on the predictor–corrector scheme should include two loops, the internal loop which would perform the iterations, and the external one, which would perform marching procedure. Define the step-size as $X = 0.1$ and the number of steps $M = 20$, which specifies the parameters of the external loop. The number of iterations needed depends on the accuracy. To this end, formulate the following *convergence criterion* for each step $(m+1)$:

$$|[y^{m+1}]_i - [y^{m+1}]_{i-1}| < \varepsilon |[y^{m+1}]_i| \tag{S11.4}$$

where i denotes the iterations and ε is a small parameter which regulates the accuracy. In the case under consideration one may set $\varepsilon = 0.001$. Once the condition (S11.4) has been met, the iterations should stop and routine turn to the next step. Note also that the first iteration employs (S10.8), (S10.6) and (S10.7), while the subsequent ones substitute (S10.5) for (S10.8). Therefore it is convenient to perform the first iteration as a separate block.

In the program below we use the notation Yx (or yx) = $y_{,x}$ and Yxx (or yxx) = $y_{,xx}$ and first state the step-size and the initial conditions (S11.3). Then we formulate the external and internal loops and use (S11.4) and while to stop the iterations when the desired accuracy is achieved:

```
> X:=0.1:
> Y[0]:=evalf(Pi/2):Yx[0]:=0:Yxx[0]:=-3.4377:
> #external loop begins
> for m from 0 to 20 do
> #internal loop begins
> yx[m+1,1]:=Yx[m]+Yxx[m]*X:
> y[m+1,1]:=Y[m]+(X/2)*Yx[m]+(X/2)*yx[m+1,1]:
> yxx[m+1,1]:=-3.4377*sin(y[m+1,1]):
> #preparing the condition (S11.4)
> s[m,1]:=1: y[m+1,1]:=1:
> for i from 2 to 10 while s[m,i-1]>0.001*y[m+1,i-1] do
> yx[m+1,i]:=Yx[m]+(X/2)*Yxx[m]+(X/2)*yxx[m+1,i-1]:
> y[m+1,i]:=Y[m]+(X/2)*Yx[m]+(X/2)*yx[m+1,i]:
> yxx[m+1,i]:=-3.4377*sin(y[m+1,i]):
> a:=y[m+1,i]:b:=yx[m+1,i]:c:=yxx[m+1,i]:
> #computing the condition (S11.4)
> s[m,i]:=abs(y[m+1,i]-y[m+1,i-1]):od:
> Y[m+1]:=a:Yx[m+1]:=b:Yxx[m+1]:=c:
```

214 FINITE DIFFERENCE METHODS

```
> x[m]:=m*X:
> od:
```

It remains to plot the results

```
> plot([[x[ii],Y[ii]]$ ii=0..20],x=0..2);
```

The graph in Fig. S4 indeed shows that near $x = 1$, which is the quarter-period, $y(x)$ takes on zero value, in agreement with the introductory comments.

This program applies to other equations as well, provided the right-hand side of (S10.7) (the g-function) and initial conditions are properly changed. Consider the equation

$$MY_{,xx} + CY_{,x} + K(Y + \alpha Y^3) = 0 \qquad (S11.5)$$

with the initial conditions

$$Y(x=0) = 0, \qquad Y_{,x}(x=0) = 10 \, \text{in/s} \qquad (S11.6)$$

which may describe a damped non-linear system. For the values of parameters $M = 100 \, \text{lb-s}^2/\text{in} = 175.12 \, \text{kg}$, $K = 70050 \, \text{N/m}$, $\alpha = 2 \, \text{in}^{-2} = 0.31 \times 10^4 \, \text{m}^{-2}$, and $C = 0$ the above equation reduces to

$$Y_{,xx} = -4(Y + 2Y^3) \qquad (S11.7)$$

and yields the following initial condition

$$Y_{,xx}(x=0) = 0 \qquad (S11.8)$$

Introducing the proper changes in the program above (these are underlined below) and specifying the step-size as $X = 0.025$, we get

```
X:=0.025:
Y[0]:=0:Yx[0]:=10:Yxx[0]:=0:
```

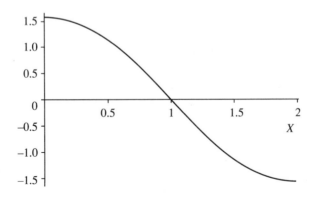

S4 To non-linear vibrations

```
for m from 0 to 20 do
yx[m+1,1]:=Yx[m]+Yxx[m]*X:
y[m+1,1]:=Y[m]+(X/2)*Yx[m]+(X/2)*yx[m+1,1]:
yxx[m+1,1]:=-4*(y[m+1,1]+2*y[m+1,1]^3):
s[m,1]:=1: y[m+1,1]:=1:
for i from 2 to 10 while s[m,i-1]>0.001*abs(y[m+1,i-1]) do
yx[m+1,i]:=Yx[m]+(X/2)*Yxx[m]+(X/2)*yxx[m+1,i-1]:
y[m+1,i]:=Y[m]+(X/2)*Yx[m]+(X/2)*yx[m+1,i]:
yxx[m+1,i]:=-4*(y[m+1,i]+2*y[m+1,i]^3):
a:=y[m+1,i]:b:=yx[m+1,i]:c:=yxx[m+1,i]:
s[m,i]:=abs(y[m+1,i]-y[m+1,i-1]):od:
Y[m+1]:=a:Yx[m+1]:=b:Yxx[m+1]:=c:
x[m]:=m*X:
od:
plot([[x[ii],Y[ii]]$ ii=0..20],x=0..0.5);
```

and produce the graph of Fig. S5.

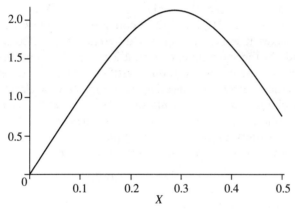

S5 To non-linear vibrations

S12. Newmark's method

(MAPLE file vfem64)

At each step m a relation between the first and second derivatives may be approximated either by

$$y_{,xx}^{m+1} = (y_{,x}^{m+1} - y_{,x}^{m})/X \tag{S12.1}$$

or by

$$y_{,xx}^m = (y_{,x}^{m+1} - y_{,x}^m)/X \tag{S12.2}$$

which, in a sense, are two extremities. On average, a smoother and better representation may be given by the 'mixture rule' of both (S12.1) and (S12.2). This leads to the following relation:

$$y_{,x}^{m+1} = y_{,x}^m + X[(1-\beta)y_{,xx}^m + \beta y_{,xx}^{m+1}] \tag{S12.3}$$

where the β-value must be specified by a user. This expression reduces to (S12.1) for $\beta = 1$ and to (S12.2) for $\beta = 0$.

The next relation of interest may be considered as motivated by the well-known result for a particle motion with constant acceleration a:

$$s = s_0 + v_0 t + at^2/2 \tag{S12.4}$$

where s is the distance and v is the velocity. Adapting this to the realm of finite difference schemes and introducing once again an adjusting parameter, set

$$y^{m+1} = y^m + y_{,x}^m X + [(1/2 - \alpha)y_{,xx}^m + \alpha y_{,xx}^{m+1}]X^2 \tag{S12.5}$$

where the α-value is at the user disposal.

The two equations (S12.3) and (S12.5) constitute *Newmark's acceleration method* which makes it easier to adjust a difference scheme to a particular problem at hand. At the same time, one should be careful about the choice of α- and β-values, as these may introduce artificial effects. For example, if $\beta < 1/2$, the method introduces negative damping. Nevertheless, an experienced user may choose these coefficients so as to improve the accuracy.

The above equations contain other schemes as particular cases and may be implemented in a variety of ways. For example, if $\alpha = 0$, (S12.5) yields the explicit result known as the *constant-acceleration method*,

$$y^{m+1} = y^m + y_{,x}^m X + 1/2 y_{,xx}^m X^2 \tag{S12.6}$$

if $\alpha = 1/4$, it yields the *average-acceleration method*,

$$y^{m+1} = y^m + y_{,x}^m X + 1/4(y_{,xx}^m + y_{,xx}^{m+1})X^2 \tag{S12.7}$$

and, eventually, if $\alpha = 1/6$, it yields the *linear-acceleration method*,

$$y^{m+1} = y^m + y_{,x}^m X + 1/6(2y_{,xx}^m + y_{,xx}^{m+1})X^2 \tag{S12.8}$$

Together with a discrete form of the differential equation to be solved,

$$Y_{,xx} = g(Y, Y_{,x}, x) \tag{S10.1}$$

(S12.3) and (S12.5) provide a complete algorithm, which applies to a system of equations too. Consider again the problem of free non-linear vibrations given by (S11.2) and (S11.3) and assume $\alpha = 0$ and $\beta = 0.5$. The program below, which employs notations of the previous section, makes use of (S12.3), (S12.5) and (S11.2), (S11.3) and yields the results for the angle of rotation similar to those obtained earlier by the predictor–corrector method,

```
> X:=0.1:y[0]:=evalf(Pi/2):yx[0]:=0:yxx[0]:=-3.4377:
  alpha:=0:beta:=0.5:
> for m from 0 to 20 do
> y[m+1]:=y[m]+yx[m]*X
  +((1/2-alpha)*yxx[m]+alpha*yxx[m+1])*X^2:
> yxx[m]:=-3.4377*sin(y[m]):yxx[m+1]:=-3.4377*sin(y[m+1]):
> yx[m+1]:=yx[m]+X*((1-beta)*yxx[m]+beta*yxx[m+1]):
> x[m]:=m*X:
> od:
> plot([[x[ii],y[ii]]$ ii=0..20],x=0..2);
```

S13. Finite difference equations

As follows from the above considerations, applications of finite difference techniques involve expressions of the type

$$\phi = \phi(y_m, y_{m+1}, \ldots, y_{m+i}) \tag{S13.1}$$

It is therefore useful to consider a closely related subject, namely, solutions to an ith order homogeneous difference equation

$$y_{m+i} + b_1 y_{m+i-1} + b_2 y_{m+i-2} + \cdots + b_i y_m = 0 \tag{S13.2}$$

with constant coefficients b_n, $n = 1, 2, \ldots, i$. The theory of these equations, which is similar to that of differential equations, assumes a fundamental solution in the form of $y_m = \lambda^m$. In order to incorporate the initial conditions $y_0, y_1, y_2, \ldots, y_{i-1}$, the general solution to (S13.2) must contain i arbitrary constants, say D_n, $n = 1, 2, \ldots, i$. One is led to the following form:

$$y_m = D_1 \lambda_1^m + D_2 \lambda_2^m + \cdots + D_i \lambda_i^m \tag{S13.3}$$

Here $\lambda_1, \lambda_2, \ldots, \lambda_i$ are the different roots of the characteristic equation

$$\lambda^{m+i} + b_1 \lambda^{m+i-1} + \cdots + b_i \lambda^m = 0 \tag{S13.4}$$

which easily follows from (S13.2) upon the above substitution $y_m = \lambda^m$.

Consider the model equation

$$y_{m+2} - 3y_{m+1} + 2y_m = 0 \tag{S13.5}$$

whose characteristic equation is

$$\lambda^{m+2} - 3\lambda^{m+1} + 2\lambda^m = 0 \tag{S13.6}$$

Assuming a non-trivial solution, $\lambda^m \neq 0$, we get $\lambda_1 = 1$ and $\lambda_2 = 2$. This yields the general solution (S13.3):

$$y_m = D_1 1^m + D_2 2^m \tag{S13.7}$$

218 FINITE DIFFERENCE METHODS

If the initial conditions are, say, $y_0 = 0$ and $y_1 = 1$, then
$$D_1 + D_2 = 0$$
$$D_1 + 2D_2 = 1 \qquad \text{(S13.8)}$$
which specifies the D-coefficients as $D_1 = -1$ and $D_2 = 1$. Eventually, the general solution (S13.7) gets the form
$$y_m = -1 + 2^m \qquad \text{(S13.9)}$$
This may be easily verified by substituting (S13.9) into (S13.5) to obtain

```
> simplify(-1+2^(m+2)-3*(-1+2^(m+1))+2*(-1+2^{\,m}));

      0
```

The case of identical (multiple) roots is different from the above one and needs special consideration, as in the theory of differential equations. Consider a difference equation
$$ay_m + by_{m+1} + cy_{m+2} = 0 \qquad \text{(S13.10)}$$
with the constant coefficients a, b and c. The characteristic equation is
$$a\lambda^m + b\lambda^{m+1} + c\lambda^{m+2} = 0 \qquad \text{(S13.11)}$$
with the roots λ_1 and λ_2 and the general solution given by
$$y_m = D_1 \lambda_1^m + D_2 \lambda_2^m \qquad \text{(S13.12)}$$
where D_1 and D_2 must be specified from initial or other conditions. Assuming that y_0 and y_1 are prescribed, set two simultaneous equations
$$y_0 = D_1 + D_2$$
$$y_1 = D_1 \lambda_1 + D_2 \lambda_2 \qquad \text{(S13.13)}$$
and solve these for the values of D_1 and D_2:

```
> eqs:={D1+D2=y[0],D1*lambda[1]+D2*lambda[2]=y[1]}:
  solve(eqs,{D1,D2});
```
$$\left\{ D2 = \frac{\text{lambda}[1]\, y[0] - y[1]}{\text{lambda}[1] - \text{lambda}[2]},\ D1 = -\frac{-y[1] + y[0]\, \text{lambda}[2]}{\text{lambda}[1] - \text{lambda}[2]} \right\}$$

It is seen that the case $\lambda_1 = \lambda_2$ indeed needs a special treatment, as the solution breaks down. Set the general solution for this case as
$$y_m = D_1 \lambda_1^m + D_2 z_m \lambda_1^m \qquad \text{(S13.14)}$$
which resembles the approach adopted in the theory of differential equations. In order to find z_m substitute the last expression in (S13.10) and get
$$D_1 \lambda_1^m (a + b\lambda_1 + c\lambda_1^2) + D_2 \lambda_1^m (az_m + bz_{m+1}\lambda_1 + cz_{m+2}\lambda_1^2) = 0 \qquad \text{(S13.15)}$$

Because of the characteristic equation (S13.11), this simplifies to

$$az_m + bz_{m+1}\lambda_1 + cz_{m+2}\lambda_1^2 = 0 \qquad (S13.16)$$

Further, (S13.11) indicates that $a = c\lambda_1^2$ and $b = -2c\lambda_1$. Substituting this in the above expression, we get

$$c\lambda_1^2 z_m - 2c\lambda_1^2 z_{m+1} + cz_{m+2}\lambda_1^2 = z_m - 2z_{m+1} + z_{m+2} = 0 \qquad (S13.17)$$

This merely shows that the difference between adjacent values of z_m remains constant

$$z_{m+1} - z_m = z_{m+2} - z_{m+1} \qquad (S13.18)$$

and, consequently, the solution is an arithmetic progression, say $z_m = m$. The general solution (S13.14) takes its final form:

$$y_m = D_1\lambda_1^m + D_2 m\lambda_1^m \qquad (S13.19)$$

S14. Stability

The term *stability* may appear in the scientific literature in various contexts and therefore needs a precise definition. To begin, it may be stated that a solution is *unstable* if the errors (truncation or round-off errors) accumulate during the computational process and may eventually become infinitely large.

Some equations exhibit this phenomenon regardless of the solution method and are said to be *inherently unstable*. This is illustrated by the simple equation of the type (S3.5)

$$F_{,x}(x) = x + F(x) \qquad (S14.1)$$

which has the exact solution as follows:

```
> eq:=diff(F(x),x)-x-F(x);
```

$$eq := \left(\frac{d}{dx}F(x)\right) - x - F(x)$$

```
> dsolve(eq,F(x));
```

$$F(x) = -x - 1 + \exp(x)\_C1$$

For the initial condition $F(x = 0) = -1$ the exponential term vanishes and the solution simplifies to

$$F(x) = -1 - x \qquad (S14.2)$$

However, a very small change in the initial condition, say, $F(x = 0) = 0.9999$, would cause a dramatic change in the solution as the exponential term no longer vanishes and in fact becomes dominant for sufficiently large x. Attempting to solve this equation by a difference scheme would result in a completely erroneous solution, as, even though for the first step the initial condition is free of error, there will be truncation and round-off errors for the subsequent steps. Thus, this solution would never have a satisfactory accuracy.

Another form of instability is associated with a particular solution algorithm and may be referred to as *partial* or *numerical instability*. Depending on the difference scheme, it may come into play even for equations which do not display inherent instability.

Consider again the model equation (S2.1)

$$F_{,x}(x) + bF(x) = 0 \tag{S14.3}$$

where b is in general a complex quantity, and apply the forward operator Δ_1 (see S1.2a) to get (S2.5):

$$f_m - f_{m-1}(1 - bX) = 0 \tag{S14.4}$$

This approach is also known as the explicit Euler method. The characteristic equation is

$$\lambda^m + b_1 \lambda^{m-1} = 0 \tag{S14.5}$$

with $b_1 = -1 + bX$, and the general solution is

$$f_m = D_1 \lambda_1^m + D_2 \lambda_2^m \tag{S14.6}$$

with λ_1 and λ_2 being the different roots of (S14.5). The coefficients D_1 and D_2 may follow from the initial conditions. It is seen that, generally speaking, the solution f_m becomes unbounded with $m \to \infty$, unless both of the roots obey the restriction $|\lambda_1| \leq 1$ and $|\lambda_2| \leq 1$. The scheme, which shows unbounded solutions *regardless* of the phenomenon described, is obviously deficient and should not be employed. On the other hand, the scheme is stable provided the above restriction is met.

In the case at hand $\lambda_1 = 0$ and $\lambda_2 = -b_1$, which yields

$$|bX - 1| < 1 \tag{S14.7}$$

as the only condition of stability. This obviously imposes a restriction on the values of the step size X. Note that, generally speaking, the quantities involved, like b and $F(x)$, may be complex. In the complex plane the expression (S14.7) formulates the so-called *region of stability*, which lies inside the unit circle with the centre at $\text{Im}(bX) = 0$, $\text{Re}(bX) = 1$.

Now we apply to (S14.3) the backward Δ_2-operator (see S1.2b) and get the difference equation

$$f_m(1 + bX) - f_{m-1} = 0 \tag{S14.8}$$

with the characteristic equation

$$\lambda^m(1 + bX) - \lambda^{m-1} = 0 \tag{S14.9}$$

This is known as the implicit Euler method. The general solution is still given by (S14.6) and the above condition of stability holds. The roots of (S14.9) are

$$\lambda_1 = 0, \quad \lambda_2 = 1/(1+bX) \tag{S14.10}$$

This gives

$$|1/(1+bX)| < 1 \tag{S14.11}$$

as the condition of stability which is less restrictive with respect to the X-value than (S14.7). The region of stability lies this time outside of the unit circle with centre at $\text{Im}(bX) = 0$, $\text{Re}(bX) = -1$.

Eventually, consider the Δ_3-operator (see S1.2c), which yields

$$f_{m+1} - f_{m-1} + 2bXf_m = 0 \tag{S14.12}$$

with the general solution given by

$$f_m = D_1\lambda_1^m + D_2\lambda_2^m + D_3\lambda_3^m \tag{S14.13}$$

The characteristic equation is

$$\lambda^{m+1} - \lambda^{m-1} + 2bX\lambda^m = 0 \tag{S14.14}$$

which provides $\lambda^m = 0$ and $(\lambda - 1/\lambda + 2bX) = 0$. The last equation has the following two roots:

$$-bX + (b^2X^2 + 1)^{1/2}, \quad -bX - (b^2X^2 + 1)^{1/2}$$

The absolute value of the last expression is greater then 1 for real b and thus this scheme may be unstable. A point to note is that under additional constraints, such as restrictions placed on the values taken on by b and X, a scheme can provide satisfactory results. Similar considerations enable one to investigate the stability of other difference schemes applied to various differential equations.

From the viewpoint of computational efficiency it is desirable to use the largest step-size possible. Often the local error limitation dictates the choice and the stability constraint is not active. For some equations, usually those possessing distinct time (or space) scales, the accuracy constraint permits a step-size which lies outside of the stability region. For these so-called *stiff equations* stability considerations control the step-size.

The interrelated issues of *accuracy* (how closely f_m approximates $F(x_m)$), *stability* (whether f_m is unbounded as $m \to \infty$) and *convergence* (whether the finite difference solution approaches the exact one in the limit) are central for the analysis of the finite-difference techniques. Later we discuss also the issue of *consistency*.

S15. Solving non-linear equations

(MAPLE file vfem65)

Though implicit schemes in general are preferable over explicit schemes from the viewpoint of their accuracy and stability, they need a special algorithm to solve for y_{m+1}. This may represent an obstacle, particularly, in the case of nonlinear systems. Below we consider two approaches to solving implicit equations of the type

$$y_{m+1} = G(y_{m+1}) \tag{S15.1}$$

The first approach, which is known as *functional iteration*, states that under some regularity conditions the sequence given by

$$y_{m+1}^{(i)} = G(y_{m+1})^{(i-1)}, \qquad i = 1, 2, \ldots \tag{S15.2}$$

converges to y_{m+1}, if the initial guess $y_{m+1}^{(0)}$ is sufficiently close to y_{m+1}. This algorithm is easy to program. An explicit scheme may provide values for an initial guess. The selected step-size must be small enough to ensure convergence.

The other approach is based on the *Newton–Raphson method*, which is applied below to a system of equations:

$$\begin{aligned} f_1(x_1, x_2, \ldots, x_n) &= 0 \\ f_2(x_1, x_2, \ldots, x_n) &= 0 \\ &\cdots \\ f_n(x_1, x_2, \ldots, x_n) &= 0 \end{aligned} \tag{S15.3}$$

For convenience, denote $\mathbf{x} = [x_1, x_2, \ldots, x_n]^T$, $\mathbf{f}(\mathbf{x}) = [f_1(\mathbf{x}), f_2(\mathbf{x}), \ldots, f_n(\mathbf{x})]^T$, and let $\alpha = [\alpha_1, \alpha_2, \ldots, \alpha_n]^T$ be a solution to (S15.3), $f_i(\alpha) = 0, i = 1, 2, \ldots, n$. Next, introduce the derivatives

$$f_{ij}(\mathbf{x}) = \partial f_i(\mathbf{x})/\partial x_j \tag{S15.4}$$

and the associated n by n matrix

$$\boldsymbol{\phi}(\mathbf{x}) = (f_{ij}(\mathbf{x})), \qquad i = 1, 2, \ldots, n; \quad j = 1, 2, \ldots, n \tag{S15.5}$$

Note that these definitions specify the *Jacobian* of the system (S15.3) as $\det(\boldsymbol{\phi})$.

Assuming the initial value of the solution to (S15.3) to be given by $\mathbf{x}_0 = [x_{10}, x_{20}, \ldots, x_{n0}]^T$ and be sufficiently close to $\alpha = [\alpha_1, \alpha_2, \ldots, \alpha_n]^T$, perform the iterative procedure by

$$\mathbf{x}_{k+1} = \mathbf{x}_k + \boldsymbol{\delta}_k \tag{S15.6}$$

where $\boldsymbol{\delta}_k$ is the solution to the set of simultaneous linear equations

$$\boldsymbol{\phi}(\mathbf{x}_k)\boldsymbol{\delta}_k = -\mathbf{f}(\mathbf{x}_k) \tag{S15.7}$$

There is a fundamental theorem which proves that $\mathbf{x}_k \to \alpha$ with $k \to \infty$, if the components of $\boldsymbol{\phi}(\mathbf{x})$ are continuous in the vicinity of a point α and $\det(\boldsymbol{\phi}(\alpha)) \neq 0$. Note once again that the initial value $\mathbf{x}_0$ must be sufficiently close to α to ensure the convergence.

SOLVING NON-LINEAR EQUATIONS   223

Consider the following example of two non-linear equations:

$$f_1(x_1, x_2) = \sin(x_1 x_2)/2 - x_2/(4\pi) - x_1/2 = 0$$
$$f_2(x_1, x_2) = [1 - 1/(4\pi)][\exp(2x_1) - e] + ex_2/\pi - 2ex_1 = 0 \qquad (S15.8)$$

for which the initial evaluation of the roots provides $x_1 = 0.4$ and $x_2 = 3.0$.

Going over to programming, denote the values of x_1 and x_2 obtained by the kth iteration as x[1,k] and x[2,k], respectively, and state (S15.8):

> f[1]:=sin(x[1,k]*x[2,k])/2-x[2,k]/(4*Pi)-x[1,k]/2;

$$f[1] := 1/2 \sin(x[1, k] x[2, k]) - 1/4 \frac{x[2, k]}{Pi} - 1/2\, x[1, k]$$

> f[2]:=(1-1/(4*Pi))*(exp(2*x[1,k])-E)
  +x[2,k]*E/Pi-2*E*x[1,k];

$$f[2] := \left(1 - \frac{1}{4\,Pi}\right)(\exp(2\, x[1, k]) - E) + \frac{x[2,\ k]\, E}{Pi} - 2\, E\, x[1, k]$$

Next, compute the derivatives (S15.4):

> for i to 2 do for j to 2 do F[i,j]:=diff(f[i],x[j,k]) od od:

and state the system of equations (S15.7):

> eq1:=F[1,1]*delta[1,k]+F[1,2]*delta[2,k]=-f[1]:
> eq2:=F[2,1]*delta[1,k]+F[2,2]*delta[2,k]=-f[2]:

Preparing the iterations, specify the above initial values for the roots:

> x[1,0]:=0.40:x[2,0]:=3.00:x[1,-1]:=100:x[2,-1]:=100:

where the last two statements merely serve to start the do-command below:

> for k from 0 to 10 while abs(x[1,k]-x[1,k-
  1])>0.0001*abs(x[1,k-1])
> and abs(x[2,k]-x[2,k-1])>0.0001*abs(x[2,k-1]) do
> s:=solve({eq1,eq2},{delta[1,k],delta[2,k]}):
> assign(s):
> x[1,k+1]:=x[1,k]+delta[1,k]:x[2,k+1]:=x[2,k]
  +delta[2,k]:od:

Here the while-statement formulates the convergence criterion with the maximal number of iterations set at 10.

To observe the solution and the number of iterations made one may key in

> x[1,k];

   -.2605992900

> x[2,k];

.6225308966

> k;

5

S16. Boundary-value problems

So far we have dealt with initial-value problems (IVP) of ordinary differential equations. Boundary-value problems (BVP) are of interest for engineering to at least the same degree, though their solution may be considerably more involved. Finite element or boundary element methods might have certain advantages over finite difference methods from this point of view.

Nevertheless, a finite difference scheme may be successfully applied to resolve a BVP, for which various techniques are available. First, sometimes a BVP may be reformulated as an IVP by a proper change of variables. Second, a solution may follow from a straightforward substitution of the boundary values into a system of equations obtained by a difference scheme. Third, the so-called *shooting approach* may transform a BVP to the associated IVP amenable to a solution by, say, iterative methods. This approach usually involves a *trial-and-error* procedure. One first ignores some of the boundary conditions and resolves an initial-value problem *assuming* missing initial conditions. The next step includes verification of whether the ignored boundary conditions have been met. The process goes on until the assumed initial conditions help to produce a solution which agrees with all the imposed boundary conditions. In doing so, one may apply various interpolation techniques to shorten the process.

The capability of performing large analytic computations provided by computer algebra may be particularly helpful for facilitating the implementation of this method. As an example, consider an equation of the second order:

$$s(Y, x, Y_x, Y_{xx}) = 0 \qquad (S16.1)$$

subjected to the boundary conditions $Y(x = a) = Y_a$ and $Y(x = b) = Y_b$. This equation may be usually rewritten as a system of two equations of the first order and resolved by one of the preceding methods for the initial conditions $Y(x = a) = Y_a$ and $Y_x(x = a) = C$ where C is considered unknown. Computing $Y(x = b) = Y(C)$ one may then solve the equation

$$Y(C) = Y_b \qquad (S16.2)$$

for the unknown C, and thereby avoid a trial-and-error procedure.

We consider below various aspects of application of the finite difference method to solution of partial differential equations.

S17. Discretization of PDEs

Various finite difference schemes for partial differential equations may be constructed. Consider the equation

$$F_{,t} = F_{,xx} \tag{S17.1}$$

where $F = F(x, t)$. Selecting, for example, the forward difference Δ_1 to represent the first time derivative and the central difference Δ_3 to represent the second spatial derivative, we get

$$(f_m^{n+1} - f_m^n)/K \approx F_{,t}$$
$$(f_{m+1}^n - 2f_m^n + f_{m-1}^n)/H^2 \approx F_{,xx} \tag{S17.2}$$

where $K = \Delta t$ and $H = \Delta x$ are the step-sizes, $t_n = nK$ and $x_m = mH$ are the coordinates of the mesh points. Substituting (S17.2) in (S17.1) yields

$$f_m^{n+1} = Rf_{m+1}^n + (1 - 2R)f_m^n + Rf_{m-1}^n \tag{S17.3}$$

with $R = K/H^2$.

Finding a solution to a partial differential equation is of course a more complicated task than a solution to an ordinary differential equation and *a priori* knowledge of its overall properties is even more essential. Therefore, the classification of PDEs is relevant for choosing an appropriate finite difference technique. Note that the general form of a second order equation may be given by

$$AF_{,xx} + 2BF_{,xy} + CF_{,yy} + DF_{,x} + EF_{,y} + UF + G = 0 \tag{S17.4}$$

with $d = -AC + B^2$ as the discriminant. The equation is *hyperbolic* if $d > 0$, *parabolic* if $d = 0$, and *elliptic* if $d < 0$. Elliptic equations are encountered in problems of elastic equilibrium or potential flow problems, parabolic equations in heat conduction and diffusion and, finally, hyperbolic equations are typical of wave propagation. The classical examples are as follows:

(a) Laplace (elliptic) operator

$$\Delta = \partial^2/\partial x_1^2 + \cdots + \partial^2/\partial x_n^2 \tag{S17.5}$$

(b) Diffusion (parabolic) operator

$$\partial/\partial t - \Delta \tag{S17.6}$$

(c) D'Alembert (hyperbolic) operator

$$\partial^2/\partial t^2 - \Delta \tag{S17.7}$$

Note that the above classification is of an essentially local nature. In the case of variable coefficients the type of equation may depend on the domain. In order

226 FINITE DIFFERENCE METHODS

to illustrate this effect, consider the equation

$$yF_{,xx} + 2xF_{,xy} + yF_{,yy} = 0 \tag{S17.8}$$

It is elliptic in the region $y^2 - x^2 > 0$, hyperbolic in the region $y^2 - x^2 < 0$ and parabolic along the lines $y^2 - x^2 = 0$.

S18. Explicit scheme

A model parabolic equation is

$$F_{,t} = F_{,xx} \tag{S18.1}$$

defined for $0 \le x \le 1$ and $0 < t \le T$ and subjected to the following initial and boundary conditions:

$$\begin{aligned} F(x,0) &= U(x), & 0 \le x \le 1 \\ F(0,t) &= G_0(t), & 0 < t \le T \\ F(1,t) &= G_1(t), & 0 < t \le T \end{aligned} \tag{S18.2}$$

This equation describes, for example, the temperature distribution in an isolated bar with the initial temperature given by the function $U(x)$ and the ends maintained at temperatures specified by the functions $G_0(t)$ and $G_1(t)$, respectively.

Fig. S6 shows a grid for the region of interest ($0 \le x \le 1, 0 < t \le T$) with the spacings $\Delta x = H = 1/M$ and $\Delta t = K = 1/N$ where M and N are arbitrary integers and $t_n = nK$ and $x_m = mH$.

Adopting the difference scheme given by (S17.3), we set the following discretization for (S18.1):

$$f_m^{n+1} = Rf_{m+1}^n + (1 - 2R)f_m^n + Rf_{m-1}^n \tag{S17.3}$$

where the subscript and superscript denote the x- and t-coordinate of the grid

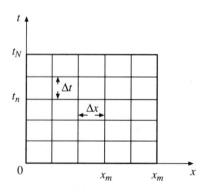

S6 Grid for a parabolic equation

point, respectively, and $R = K/H^2$. Fig. S7 shows the template of this explicit scheme (points involved in (S17.3)) with the time difference denoted by a solid square and space difference by a hollow square.

The above equation (S17.3) applies for $1 \leq m \leq M - 1$ only, while the boundary conditions $G_0(t)$ and $G_1(t)$ control the f-values for $m = 0$ and for $m = M$. The initial condition given by $U(x)$ specifies the f-value for any m when $n = 0$. Equation (S17.3) enables one to explicitly calculate f_m^{n+1} by the time-step procedure from the f-values at the earlier time t_n. Note that all these values of f at any time level must be calculated before advancing to the next time instant.

The corners $(0, 0)$ and $(1, 0)$ may represent a difficulty from the viewpoint of initial and boundary conditions, if these do not match one another. In this case $F(x, t)$ would suffer a jump at the corners. Usually, $f(0, 0)$ is prescribed as the average of the two limits, $\lim U(x)$ as $x \to 0$ and $\lim G_0(t)$ as $t \to 0$. Another possibility is to comply with one of the limits only, which would also introduce an error.

As an example, consider the case $U(x) = 0$, $G_0(t) = G_1(t) = 100$ and specify the step-sizes as $H = 0.2$ and $K = 0.01$. This yields $R = K/H^2 = 1/4$. For $t = 0$ ($n = 0$) the f-values may be specified as zero to comply with the initial condition, for $t = 0.2$ ($n = 1$) let $f_0^1 = f_5^1 = 100$ and $f_1^1 = f_2^1 = f_3^1 = f_4^1 = 0$ to enforce the boundary conditions (see Section S23 for more details about the treatment of boundary conditions). Then (S17.3) yields, for $t_2 = 0.4$ ($n = 2$),

$$f_m^2 = (1/4 f_{m+1} + 1/2 f_m + 1/4 f_{m-1})^1 \tag{S18.3}$$

where $1 \leq m \leq M - 1$. This gives $f_1^{(2)} = f_4^{(2)} = 25$ and zero for the rest of the internal points at $t = 0.4$. Now one can go over to the next time-step, etc., and observe the gradual heat diffusion into the bar. Section S24 presents an example in more detail.

It can be shown that this finite difference solution converges to the exact one if the spacings K and H are small enough and $0 < R \leq 1/2$. In general, the scheme is sensitive to the choice of R which controls its performance. A further discussion of this and related subjects is given in Section S20.

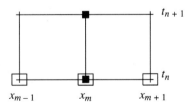

S7 Template for the explicit scheme

S19. Implicit scheme

Implicit schemes provide, figuratively speaking, a kind of feedback, and lead to a system of equations in which the unknown values affect each other in a more balanced way. They usually offer a better performance then explicit ones at the expense of a more elaborate treatment. In order to construct an implicit scheme for (S18.1) represent F_{xx} by a form evaluated at the advanced instant of time t_{n+1}, instead of t_n as in Section S18.

In the notation of the previous section we get

$$(f_m^{n+1} - f_m^n)/K = (f_{m-1}^{n+1} - 2f_m^{n+1} + f_{m+1}^{n+1})/H^2 \quad (S19.1)$$

where $K = \Delta t$ and $H = \Delta x$. This yields

$$f_m^n = -Rf_{m-1}^{n+1} + (1 + 2R)f_m^{n+1} - Rf_{m+1}^{n+1} \quad (S19.2)$$

with $R = K/H^2$. Fig. S8 shows the template for this implicit scheme with the time difference denoted by a solid square and space difference by a hollow square. Comparing this figure with Fig. S7, one may appreciate better the difference between the explicit and implicit schemes.

The boundary and initial conditions are still given by (S18.2).

At any time instant, (S19.2) applies for each point $1 \leq m \leq M - 1$ which delivers a system of $M - 1$ simultaneous equations governing $M - 1$ unknown values f_m^{n+1}. Here $M = 1/H$. Equations (S18.2) specify the rest of the values.

As an example, consider again the case $U(x) = 0$, $G_0(t) = G_1(t) = 100$ and specify the step-sizes as, say, $H = 0.2$ and $K = 0.04$, which yields $R = 1$. For $t = 0$ $(n = 0)$ get $f_0^0 = f_5^0 = 100$ and $f_1^0 = f_2^0 = f_3^0 = f_4^0 = 0$ to enforce the boundary conditions. Then (S19.2) yields, for $t_1 = 0.2$ $(n = 1)$,

$$f_m^0 = (-f_{m-1} + 3f_m - f_{m+1})^1 \quad (S19.3)$$

where $1 \leq m \leq M - 1$. After resolving this system of simultaneous equations, one can go over to the next time-step, etc. Section S23 contains more details about the treatment of boundary conditions and Section S25 an elaborated example.

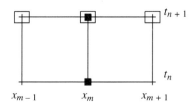

S8 Template for the implicit scheme

It can be shown that this finite difference solution converges to the exact one regardless of the R-value, if the spacings K and H are small enough. Thus, unlike the explicit scheme, the implicit one does not put a restriction on R, which may be a substantial advantage for problems extending over large values of time. A further discussion of this and related subjects is given in Section S20.

S20. Stability

Consider first the explicit scheme (S17.3)

$$f_m^{n+1} = R f_{m+1}^n + (1 - 2R) f_m^n + R f_{m-1}^n \tag{S17.3}$$

and assume that f_m^0, which specifies the initial condition, is given as the zig-zag alternating peaks from $-\sigma$ to σ, in other words, $f_m^0 = (-1)^{m+1}\sigma$. Equation (S17.3) yields

$$f_m^1 = (1 - 4R) f_m^0 \tag{S20.1}$$

and

$$|f_m^n| = |(1 - 4R)^n f_m^0| = |\sigma(1 - 4R)^n| \tag{S20.2}$$

The last relation shows that for $R > 1/2$ the explicit scheme magnifies the peaks without limit as $n \to \infty$, which is typical of unstable behaviour. For $R < 1/2$ the scheme smears the peaks out as $n \to \infty$ and is therefore stable. This is consistent with the remarks made at the end of Section S18 and the example treated there.

There are of course more systematic methods of investigation of stability. One of them assumes a process similar to that in the above and is known as the *von Neumann stability analysis*. Let $U(x)$ specify the initial condition ($t = 0$) and $e^{(i\beta x)}$, $\beta > 0$, represent a typical term (without a constant coefficient) of its Fourier expansion. Assuming the separation of variables is valid, this term becomes $\sigma(t) e^{(i\beta x)}$ at time t. Substituting this representation in the difference scheme under consideration and considering the evolution of $\sigma(t)$, one may draw a conclusion on the stability of the scheme. Begin with the explicit scheme (S17.3) written as

$$(f_m^{n+1} - f_m^n)/K = (f_{m-1}^n - 2f_m^n + f_{m+1}^n)/H^2 \tag{S20.3}$$

and substitute $f_m^n = \sigma(t) e^{(i\beta x)}$ to obtain

$$[\sigma(t+K) e^{(i\beta x)} - \sigma(t) e^{(i\beta x)}]/K$$
$$= \sigma(t)[e^{(i\beta x - H)} - 2 e^{(i\beta x)} + e^{(i\beta x + H)}]/H^2 \tag{S20.4}$$

This yields

$$\sigma(t + K) = \sigma(t)[1 - 4R \sin^2(\beta H/2)] \tag{S20.5}$$

where $K = \Delta t$ and $H = \Delta x$. Since $\sigma(0) = 1$, this provides

$$\sigma(t) = [1 - 4R\sin^2(\beta H/2)]^{t/K} \qquad (S20.6)$$

For $\sigma(t)$ to remain bounded (as $\Delta t \to 0$ and $\Delta x \to 0$) with increasing t, one must impose the restriction

$$-1 \leq 1 - 4R\sin^2(\beta H/2) \leq 1 \qquad \text{for all} \quad \beta H \qquad (S20.7)$$

As long as $R \geq 0$, the expression of interest complies with the upper bound. In order to comply with the lower bound, set

$$R \leq 1/[2R\sin^2(\beta H/2)] \qquad (S20.8)$$

which produces two limiting relevant cases, $\beta = 0$ and $\beta = \pi/H$, and leads to the restriction $R \leq 1/2$. Thus the region of stability is $0 \leq R \leq 1/2$, in agreement with the previous considerations. Note that, because of the round-off errors, the frequency β should be viewed as arbitrary, even though the initial condition itself may put some restrictions on this value.

Going over to the implicit scheme, set

$$(f_m^{n+1} - f_m^n)/K = (f_{m-1}^{n+1} - 2f_m^{n+1} + f_{m+1}^{n+1})/H^2 \qquad (S20.9)$$

and substitute $f_m^n = \sigma(t)e^{(i\beta x)}$. This yields

$$\sigma(t+K)/\sigma(t) = 1/[1 + 4R\sin^2(\beta H/2)] \qquad (S20.10)$$

Since this value is less than or equal to 1 for all R, this implicit scheme is *unconditionally stable*.

S21. Consistency

A particular feature of the finite difference procedure applied to PDEs is that it may reproduce a solution of the equation which is different from that under consideration. To this end, consider the explicit approximation to

$$F_{,t} = F_{,xx} \qquad (S21.1)$$

proposed by DuFort and Frankel:

$$(f_m^{n+1} - f_m^{n-1})/(2K) = (f_{m-1}^n - f_m^{n-1} - f_m^{n+1} + f_{m+1}^n)/H^2 \qquad (S21.2)$$

which is unconditionally stable. Using Taylor's expansion (see Section S1) one may show that the truncation error of (S21.2) is

$$(f_{m-1}^n - f_m^{n-1} - f_m^{n+1} + f_{m+1}^n)/H^2 - (f_m^{n+1} - f_m^{n-1})/(2K)$$
$$- (F_{,xx} - F_{,t})_m^n = O(K^2) + O(H^2) + O[(K^4/H^2) - (K/H)^2 F_{,tt}]) \qquad (S21.3)$$

where $K = \Delta t$ and $H = \Delta x$. One may observe that if $K \to 0$ and $H \to 0$ so as to ensure that $(K/H)^2 = \text{const.} = c$, then the term $c^2 F_{,tt}$ remains and (S21.2) reproduces not the original equation (S21.1) but a completely different equation:

$$F_{,t} + c^2 F_{,tt} = F_{,xx} \tag{S21.4}$$

Thus, to arrive at a *consistent* representation of (S21.1) with the help of (S21.2) K must go to zero faster than H, thereby making the term $F_{,tt}$ disappear.

Another example deals with the explicit scheme of Section S18,

$$(f_m^{n+1} - f_m^n)/K = (f_{m-1}^n - 2f_m^n + f_{m+1}^n)/H^2 \tag{S21.5}$$

applied to approximate the equation (S21.1). The truncation error obtained with the help of Taylor's series is

$$(f_{m-1}^n - 2f_m^n + f_{m+1}^n)/H^2 - (f_m^{n+1} - f_m^n)/K - (F_{,xx} - F_{,t})_m^n = O(K) + O(H^2) \tag{S21.6}$$

which tends to zero as $K \to 0$ and $H \to 0$. Thus (S21.6) is consistent with the original equation (S21.1).

The convergence, stability and consistency are three main features relevant for evaluating the performance of a finite difference scheme. Though their definitions are different, the three relate to one another. Futhermore, it can be shown (Lax's equivalence theorem) that for a one-step scheme the consistency and stability imply convergence.

S22. Other FD-schemes

It is evident from the above considerations that though the equation at hand is unique, its finite difference approximation admits a variety of forms. Besides the schemes of Sections S18 and S19 and the DuFort and Frankel scheme, there exist many other approaches.

In order to illustrate some of the possibilities arising, we deal below only with two-step schemes which employ central differencing in space. In this case a generic finite difference scheme for parabolic PDEs may be cast as follows:

$$f_m^{n+1} - f_m^n = R[\epsilon(f_{m-1}^{n+1} - 2f_m^{n+1} + f_{m+1}^{n+1}) + (1 - \epsilon)(f_{m-1}^n - 2f_m^n + f_{m+1}^n)] \tag{S22.1}$$

where $0 \leq \epsilon \leq 1$ and $R = K/H^2$. The two extremities $\epsilon = 0$ and $\epsilon = 1$ recover the forward (explicit) Euler and backward (implicit) Euler methods, respectively. Setting $\epsilon = 1/2$, which amounts to the trapezoidal rule in time, one gets the Crank–Nicolson scheme. Fig. S9 shows the template for this popular

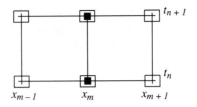

S9 Template for the Crank–Nicolson scheme

scheme with the time difference denoted by a solid square and space difference by a hollow square.

This scheme can be shown to have the truncation error (order of accuracy) of $O(\Delta t^2 + \Delta x^2)$ which is an improvement over the above Euler methods which have a truncation error of $O(\Delta t + \Delta x^2)$ (see S21.3). It is also stable for all values of R.

Indeed, the von Neumann analysis of stability (see Section S20) of the generic relation (S22.1) yields

$$\sigma(t+K) - \sigma(t) = 2\sigma(t+K)R\epsilon[\cos(\beta H) - 1] + 2\sigma(t)R(1-\epsilon)[\cos(\beta H) - 1] \tag{S22.2}$$

For the Crank–Nicolson scheme ($\epsilon = 1/2$) this becomes

$$\sigma(t+K)/\sigma(t) = [1 - 2R\sin^2(\beta H/2)]/[1 + 2R\sin^2(\beta H/2)] \tag{S22.3}$$

and shows that the scheme is stable for all $R > 0$.

By varying the value of ϵ in (S22.1) one may arrive at schemes with different degree of implicitness. Naturally, in the case of two or three spatial coordinates there exists even greater possibility for the formulation of various finite difference approaches.

S23. Treatment of boundary conditions

Consider again the one-dimensional heat equation

$$F_{,t} = F_{,xx} \tag{S18.1}$$

where $0 \leq x \leq 1$ and $t \geq 0$. The initial conditions may be formulated as

$$F(x, 0) = U(x), \qquad 0 \leq x \leq 1 \tag{S23.1}$$

and boundary conditions as

$$\begin{aligned} p_1 F_{,x} - q_1 F &= G_0(t), & x=0, \quad t \geq 0 \\ p_2 F_{,x} + q_2 F &= G_1(t), & x=1, \quad t \geq 0 \end{aligned} \tag{S23.2}$$

where $G_0(t)$ and $G_1(t)$ are bounded as $t \to \infty$ and continuous.

A particular case of (S23.1) and (S23.2), the so-called Dirichlet type boundary conditions, was considered in Sections S18 and S19. For this case $p_1 = p_2 = 0$ and $q_1 \neq q_2 \neq 0$. For the Neumann type boundary conditions $q_1 = q_2 = 0$ and $p_1 \neq p_2 \neq 0$. Of course, the mixed type may involve all these coefficients having non-zero values. These definitions generalize to the two- or three-dimensional case. In approximating the boundary conditions it is desirable to attain the same order of accuracy which characterizes the finite difference scheme for the governing equation. There are, however, cases when violation of this rule may be justified. Like the equation itself, finite difference approximation of the boundary conditions admits a variety of forms.

For example, the one-sided approach, which assumes the following relations instead of (S23.2), may be applied:

$$p_1(f_1^n - f_0^n)/H - q_1 f_0^n = G_0^n \qquad \text{(S23.3a)}$$
$$p_2(f_M^n - f_{M-1}^n)/H + q_2 f_M^n = G_1^n \qquad \text{(S23.3b)}$$

where $M = 1/H$. Using, say, (S17.3)

$$f_m^{n+1} = R f_{m+1}^n + (1 - 2R) f_m^n + R f_{m-1}^n \qquad \text{(S17.3)}$$

which applies for $m = 1, \ldots, M - 1$, one may find f_1^{n+1} and then find f_0^{n+1} from (S23.3a). A similar procedure applies to (S23.3b).

Another possibility is using the central difference for approximating (S23.2):

$$p_1(f_1^n - f_{-1}^n)/(2H) - q_1 f_0^n = G_0^n \qquad \text{(S23.4a)}$$
$$p_2(f_{M+1}^n - f_{M-1}^n)/(2H) + q_2 f_M^n = G_1^n \qquad \text{(S23.4b)}$$

where f_{-1}^n and f_{M+1}^n are located outside the domain of interest. From (S23.4a) one gets

$$f_{-1}^n = f_1^n - (2R q_1 f_0^n - 2H G_0^n)/p_1 \qquad \text{(S23.5)}$$

Equation (S17.3) yields, for $m = 0$,

$$f_0^{n+1} = R f_1^n + (1 - 2R) f_0^n + R f_{-1}^n \qquad \text{(S23.6)}$$

Equations (S23.5) and (S23.6) enable one to eliminate f_{-1}^n and get the following relation for $m = 0$:

$$f_0^{n+1} = 2R f_1^n + (1 - 2R - 2RH q_1/p_1) f_0^n - 2RH G_0^n/p_1 \qquad \text{(S23.7)}$$

Similarly, for $m = M$ one gets

$$f_M^{n+1} = 2R f_{M-1}^n + (1 - 2R + 2RH q_2/p_2) f_M^n + 2RH G_1^n/p_2 \qquad \text{(S23.8)}$$

while the above equation (S17.3) applies for the rest of the grid points, namely, for $m = 1, \ldots, M - 1$.

234 FINITE DIFFERENCE METHODS

In problems having two or three spatial dimensions, treatment of boundary conditions may be a complicated task to deal with unless a simple boundary and symmetry are present (see Section S28).

S24. Heat conduction in a slab (explicit scheme)

(MAPLE file vfem66)

Consider an infinite parallel-sided slab $(0 \leq \xi \leq L)$ initially at a uniform temperature θ_0, in other words $\theta(\tau = 0, \xi) = \theta_0$, where τ is time. Its two faces are then kept at a constant temperature θ_1, giving rise to unsteady-state heat conduction, which is governed by the equation

$$\theta_{,\xi\xi} = (1/\alpha)\theta_{,\tau} \tag{S24.1}$$

where α is the slab thermal diffusivity.

First, state this problem in dimensionless form by introducing the variables

$$F = (\theta - \theta_0)/(\theta_1 - \theta_0), \qquad t = \tau\alpha/L^2, \qquad x = \xi/L \tag{S24.2}$$

Now (S24.1) gets the familiar form of (S18.1)

$$F_{,t} = F_{,xx} \tag{S24.3}$$

The initial and boundary conditions are

$$\begin{aligned} F(t=0, 0 \leq x \leq 1) &= 0 \\ F(t>0, x=0 \text{ or } x=1) &= 1 \end{aligned} \tag{S24.4}$$

The program below makes use of the explicit scheme (S17.3):

$$f_m^{n+1} = Rf_{m+1}^n + (1-2R)f_m^n + Rf_{m-1}^n, \qquad m = 1, \ldots, M-1 \tag{S17.3}$$

and the boundary conditions following from (S23.3) and (S24.4):

$$f_0^n = 1, \qquad f_M^n = 1 \tag{S24.5}$$

State first the relevant constants and set the grid

```
> H:=1/20:K:=0.001:R:=K/H^2:M:=1/H:N:=300:
```

where N specifies the maximal number of time steps. The values H and K are so chosen as to comply with the stability criteria of Section S20. For $t = 0$ the temperature is zero all over the slab, so one may begin with the time instant $n = 1$ ($t = K$) and impose zero temperature everywhere except for the faces by the following commands:

```
> for n to N do
> [n,0]:=1:f[n,M]:=1 od:
> for m to M-1 do f[1,m]:=0 od:
```

HEAT CONDUCTION IN A SLAB (IMPLICIT SCHEME)

Next, apply (S17.3) to compute the temperature for any internal point (n, m) of the grid

```
> for n from 2 to N do
> for m to M-1 do
> f[n,m]:=R*f[n-1,m+1]+(1-2*R)*f[n-1,m]+R*f[n-1,m-1]
> od:od:
```

In order to observe the gradual diffusion of heat into the slab and its transition to a steady state, we present below the results for various x at $t = 0.1$ and $t = 0.3$. Because of the obvious symmetry, it is sufficient to consider the region $0 \le x \le 1/2$ only:

> $t = 0.1$

```
> f[100,s] $s=0..10;
```

1, .9253710686, .8525841193, .7834348260, .7196275937, .6627332155, .6141502718, .5750712034, .5464537865, .5289985244, .5231323049

$t = 0.3$

```
> f[300,s] $s=0..10;
```

1, .98953550, .9796444446, .9700947555, .9612814329, .9534214898, .9467084643, .9413076531, .9373520425, .9349390324, .9341280392

In general, computations for several values of H and K are required to draw a conclusion on the accuracy of the finite difference solution. In this case the exact solution is available. Fot $t = 0.1$ and $x = 0.5$ it yields the value 0.526 compared to the above approximate value 0.523, which shows satisfactory accuracy.

In case of large computations a special care should be taken of the memory. In order to avoid the storage of values that are no longer wanted, the above program may be properly modified so as to 'forget' the results immediately after printing (plotting) them, since (S17.3) relates only the two 'adjacent' instants of time.

S25. Heat conduction in a slab (implicit scheme)

(MAPLE file vfem67)

Consider once again the solution of the problem stated in Section S24, this time by the implicit method given by (S19.2):

$$f_m^n = -Rf_{m-1}^{n+1} + (1+2R)f_m^{n+1} - Rf_{m+1}^{n+1} \qquad \text{(S19.2)}$$

236 FINITE DIFFERENCE METHODS

while the initial and boundary conditions are still given by (S24.4)

$$F(t=0, 0 \leq x \leq 1) = 0 \tag{S24.4a}$$
$$F(t > 0, x = 0 \text{ or } x = 1) = 1 \tag{S24.4b}$$

For $m = 1$ and $m = M - 1$ (S19.2) yields, respectively,

$$f_1^n + R f_0^{n+1} = (1 + 2R) f_1^{n+1} - R f_2^{n+1} \tag{S25.1}$$

and

$$f_{M-1}^n + R f_M^{n+1} = (1 + 2R) f_{M-1}^{n+1} - R f_{M-2}^{n+1} \tag{S25.2}$$

which allow for incorporation of the boundary conditions. It is useful to specify the boundary conditions with the help of arbitrary functions $G_0(t)$ and $G_1(t)$, as in (S18.2), of which (S24.4b) is a particular case. This would allow other cases of boundary conditions to be treated by merely modifying some commands of the program. Thus, (S25.1) and (S25.2) take the form

$$f_1^n + R G_0^{n+1} = (1 + 2R) f_1^{n+1} - R f_2^{n+1} \tag{S25.3}$$

and

$$f_{M-1}^n + R G_1^{n+1} = (1 + 2R) f_{M-1}^{n+1} - R f_{M-2}^{n+1} \tag{S25.4}$$

where $G_0^n = G_0(t_n)$ and $G_1^n = G_1(t_n)$. According to (S24.4b), $G_0^n = G_1^n = 1$. Together with (S19.2) written for $2 \leq m \leq M - 2$, (S25.3) and (S25.4) provide a system of $M - 1$ simultaneous equations for $M - 1$ unknown values f_m^{n+1}. It can be shown that this sytem has a tridiagonal matrix of coefficients that greatly simplifies its solution.

The implicit method is unconditionally stable, so one may use larger stepsizes than those of the explicit method (see Section S24). State the relevant constant and initial and boundary conditions. The commands, which should be modified while reformulating boundary conditions, are underlined below:

```
> H:=1/10:K:=0.0125:R:=K/H^2:M:=1/H:N:=20:
> for n from 0 to N+1 do
> G0[n]:=1:G1[n]:=1:
>f[n,0]:=G0[n]:f[n,M]:=G1[n]: od:
> for m to M-1 do f[0,m]:=0 od:
```

Then formulate (S25.3), (S25.4) and (19.2) and solve the above-mentioned system of equations:

```
> for n from 0 to N do
> eq[1]:=(1+2*R)*f[n+1,1]-R*f[n+1,2]-f[n,1]-R*G0[n+1]:
> eq[M-1]:=(1+2*R)*f[n+1,M-1]-R*f[n+1,M-2]-f[n,M-1]
  -R*G1[n+1]:
> for m from 2 to M-2 do
```

```
> eq[m]:=-f[n,m]-R*f[n+1,m-1]+(1+2*R)*f[n+1,m]
> -R*f[n+1,m+1] od:
> s:={eq[mm] $mm=1..M-1}:
> a:=solve(s,{f[n+1,mm] $mm=1..M-1}):
> assign(a):
> od:
```

For example, for $t = 0.25$ ($n = 20$) one gets, for $0 \leq x \leq 1/2$,

```
> f[20,z] $z=0..M/2;
```

1, .9612145587, .9262258695, .8984588975, .8806315379, .8744886769

which practically coincides with the exact solution.

S26. Heat conduction in a solidifying alloy, 1

Fig. S10 shows a cylindrical mould charged with a molten alloy of two metals at an initial temperature of $T_{\text{liq}} = 400°F = 204.44°C$, which is the liquidus temperature of the alloy. One of its ends $x = 0$ is maintained at the solidus temperature of the alloy, which is $T_{\text{sol}} = 150°F = 65.55°C$, by the stream of oil. Apart from this, the mould is well insulated. These conditions give rise to an unsteady heat conduction in the alloy. The temperature distribution $T(t, x)$ where $0 \leq x \leq L$ ($L = 2\,\text{ft} = 0.6096\,\text{m}$) and $0 \leq t \leq t^*$ ($t^* = 80\,\text{h}$) is the function of interest herein.

The other relevant data are: $k = 10.1\,\text{BTU}\,\text{h}^{-1}\text{ft}^{-1}°\text{F}^{-1} = 0.1747\,\text{W/h cm}\,°\text{C}$, $\rho = 540\,\text{lb ft}^{-3} = 8649.72\,\text{kg/m}^3$, $c_\text{p} = 0.038\,\text{BTU lb}^{-1}°\text{F}^{-1} = 159.095\,\text{Ws/kg}\,°\text{C}$ for both solid and liquid. Here k is the thermal conductivity, ρ is density and c_p is specific heat. Next, the latent heat of fusion is $\Delta h = 120\,\text{BTU lb}^{-1} = 278.4\,\text{kJ/kg}$. The quantity z evaluates the amount of solid formed due to a

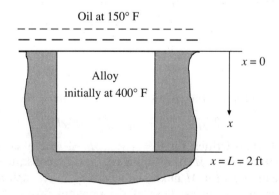

S10 Transient state of solidifying alloy

238 FINITE DIFFERENCE METHODS

small decrease in the temperature of the mixture. A simple empirical relation for $z(T)$ is given by

$$z(T) = 0.05823 - 0.000498T + 0.1484 \times 10^{-5} T^2 - 0.1483 \times 10^{-8} T^3 \quad (S26.1)$$

where T is given in °F. Analysing a heat balance on a differential element and taking into consideration a term arising from latent heat effects, one arrives at the following equation for the alloy temperature:

$$T_{,xx} = (\rho/k)[c_p + z(T)\Delta h]T_{,t} \quad (S26.2)$$

which is a non-linear PDE of the second order. This equation is subject to the initial condition

$$T(t=0, 0 \leq x \leq L) = T_{\text{liq}} \quad (S26.3)$$

and boundary conditions

$$T(t>0, x=0) = T_{\text{sol}} \quad (S26.4a)$$
$$T_{,x}(t>0, x=L) = 0 \quad (S26.4b)$$

A possible (implicit) finite difference approximation for the governing equation (S26.2) is

$$f_{m-1}^{n+1} - 2f_m^{n+1} + f_{m+1}^{n+1} = \gamma_m^n (f_m^{n+1} - f_m^n) \quad (S26.5)$$

where

$$\gamma_m^n = (\rho/k)[c_p + z_m^n \Delta h](H^2/K) \quad (S26.6)$$

and $f_m^n = T(t=t_n, x=x_m)$, $H = \Delta x$, $K = \Delta t$. Note, (S26.6) represents (S26.2) as a linearized finite-difference scheme by introducing a piecewise approximation for $z(t)$. At each point the value of z_m^n is taken as that corresponding to the previous time-step.

Since the boundary condition (S26.4b) involves the derivative of $T(t, x)$, the present problem has yet another distinct feature as compared to the problems considered in Sections S24 and S25. To this end, write Taylor's expansion

$$f_{m-1} - f_m = -F_{,x} H + (H^2/2) F_{,xx} + O(H^3) \quad (S26.7)$$

where $f_m = F(x_m)$. In view of (S26.4b) and (S26.7), we get for $x = L$

$$T_{,xx} = 2(f_{m-1} - f_m)/H^2 + O(H) \quad (S26.8)$$

Now (S26.2) provides, for the vicinity of $x = L$ (or $m = M$),

$$-2f_{M-1}^{n+1} + (2 + \gamma_M^n) f_M^{n+1} = \gamma_M^n f_M^n \quad (S26.9)$$

The equation (S26.5) written for $m = 1, \ldots, M-1$ (the internal points) and equation (S26.9) for $n = M$ (the boundary $x = L$) constitute a system of simultaneous equations for M unknown values f_m^{n+1}. At each instant of time the values of γ_m^n are taken as those of the previous instant by resorting to (S26.1).

S27. Heat conduction in a solidifying alloy, 2

(MAPLE file vfem68)

Turning to programming, specify the relevant constants given in Section S26, step-sizes H and K:

```
> rho:=540:k:=10.1:c:=0.038:h:=120:L:=2:
> K:=5:M:=10:H:=L/M:N:=16:W:=rho*H^2/(k*K):
```

where W is the factor appearing in (S26.6), and the z-dependence given by (S26.1)

```
> z:=0.05823-.000498*x+.1484*10^(-5)*x^2
  -.1483*10^(-8)*x^3:
```

This empirical function of temperature is shown in Fig. S11 for the interval of interest.

Now formulate the initial and boundary conditions according to (S26.3) and (S26.4) and the initial value of γ_m^n:

```
> for n from 0 to N+1 do f[n,0]:=150 od:
> for m to M do f[0,m]:=400:
> gamma[0,m]:=W*(c+h*subs(x=400,z)) od:
```

The following commands state the system of equations (S26.5) and (S26.9), as elaborated in the end of Section S26. After finding the solution to this system, the program calls the function $z(x)$ and finds the corresponding value of γ_m^{n+1} to be used in the next time-step:

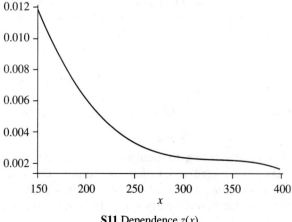

S11 Dependence $z(x)$

```
> for n from 0 to N do
> for m to M-1 do
> eq[n,m]:=f[n+1,m-1]-2*f[n+1,m]+f[n+1,m+1]
  -gamma[n,m]*(f[n+1,m]-f[n,m]) od:
> eq[n,M]:=-2*f[n+1,M-1]+(2+gamma[n,M])*f[n+1,M]
  -gamma[n,M]*f[n,M]:
> s:={eq[n,mm] $mm=1..M};
> a:=evalf(solve(s,{f[n+1,m1] $m1=1..M})):
> assign(a):
> for i to M do gamma[n+1,i]:=W*(c+h*subs(x=f[n+1,i],z)) od:
> od:
```

For example, for $t = 25\,\mathrm{h}$ ($n = 5$) one gets

```
> f[5,q] $q=0..10;
```

    150, 173.6596801, 195.5423786, 215.0461808, 232.0382520, 246.5095554,
    258.4499901, 267.8221003, 274.5716441, 278.6459855, 280.0084317

and for $t = 80\,\mathrm{h}$ ($n = 16$)

```
> f[16,q] $q=0..10;
```

    150, 158.2202367, 166.1009839, 173.3911088, 179.9154296, 185.5569458,
    190.2400982, 193.9174684, 196.5601921, 198.1515267, 198.6828539

Varying the step-sizes, one may conclude that this solution has a satisfactory accuracy. Note that in this case the non-linearity of the problem has not been an obstacle for the application of a finite difference approach.

S28. Treatment of irregular boundaries

For a sufficiently symmetrical and 'regular' domain, one may specify a grid so as to locate certain points on the boundaries which makes the satisfaction of the Dirichlet boundary conditions straightforward. In many practical situations this is not the case. Fig. S12 shows a square grid covering a domain with the irregular boundary E-A-B-C.

Assume that the equation under consideration involves the spatial second-order operator, $\nabla^2 F(x,y) = F_{,xx} + F_{,yy}$ where $F(x,y)$ is an unknown function. Further, assume that on the boundary $F(x,y)$ must satisfy the relation

$$F(x,y) = G(x,y) \quad \text{on} \quad E\text{-}A\text{-}B\text{-}C \tag{S28.1}$$

An obvious way of attempting to incorporate (S28.1) is to approximate the curved boundary by a jagged series of steps fitting in the square grid. This, which is known as the *interpolation of degree zero*, may provide a satisfactory accuracy, if the step-size is sufficiently small. In general, this technique has a limited efficiency.

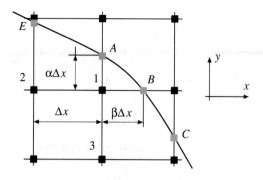

S12 Irregular boundary

Another, more elaborated, approach is based on Taylor's expansion and is known as the *interpolation of degree two*.

Referring to Fig. S12, we get

$$F_A = F_1 + \alpha\Delta x F_{,y} + (\alpha\Delta x)^2 F_{,yy}/2 + \cdots$$
$$F_3 = F_1 - \Delta x F_{,y} + (\Delta x)^2 F_{,yy}/2 + \cdots$$
$$F_B = F_1 + \beta\Delta x F_{,x} + (\beta\Delta x)^2 F_{,xx}/2 + \cdots$$
$$F_2 = F_1 - \Delta x F_{,x} + (\Delta x)^2 F_{,xx}/2 + \cdots$$
(S28.2)

where the terms of order $(\Delta x)^3$ have been neglected. This, after some algebra and elimination of the first derivatives of $F(x,y)$, yields the following finite difference approximation for $\nabla^2 F(x,y)$ valid in the vicinity of the curved boundary:

$$\nabla^2 F(x,y) = 2/(\Delta x)^2 [F_2/(\beta + 1) + F_3/(\alpha + 1) \\ + F_A/(\alpha^2 + \alpha) + F_B/(\beta^2 + \beta) - (\alpha + \beta)F_1/(\alpha\beta)] + O(\Delta x) \quad (S28.3)$$

where F_A and F_B are specified by the boundary conditions (S28.1) and F_1, F_2 and F_3 are the unknown values at the corresponding nodes of the square grid.

A similar relation may be written for other pairs of boundary points. The Neumann type of boundary conditions requires a more involved treatment and is not considered herein. Generally speaking, from the viewpoint of treatment of irregular boundaries, the finite element method offers certain advantages over the present one.

S29. Torsion of a shaft, 1

The elastic torsion of a shaft leads to the elliptic equation

$$\nabla^2 F(x,y) = -2 \qquad (S29.1)$$

242 FINITE DIFFERENCE METHODS

with the homogeneous boundary condition

$$F(x,y) = 0 \tag{S29.2}$$

on the surface of the shaft and $\nabla^2 = F_{,xx} + F_{,yy}$. Equations of the type (S29.1) are known as Poisson's equations and ∇^2 as Laplace's operator. The angle of twist θ per unit length is then given by

$$\theta = M/(2GQ) \tag{S29.3}$$

where M is the torque, G is the shear modulus and Q is

$$Q = \int_S F \, ds \tag{S29.4}$$

where S is the cross-sectional area.

In order to compare the results of a finite difference method with the exact ones, assume the cross-section of the shaft to be an ellipse, in which case the exact result is given by

$$F = (1 - x^2/a^2 - y^2/b^2)[a^2 b^2/(a^2 + b^2)] \tag{S29.5}$$

where a and b are the major and minor semi-axes shown in Fig. S13.

Even though this problem is treated below with the help of a square grid

$$\Delta x = \Delta y = a/n = b/m \tag{S29.6}$$

with integers n and m, it enables one to study particular issues due to the presence of the curved boundary. It also illustrates rational ways of accounting for the symmetry of the problem, because of which one may consider a quadrant of the ellipse only, as shown in Fig. S14.

A basic approximation for (S29.1) follows from the central difference operator (S1.3) as

$$(f_{i-1,j} - 2f_{i,j} + f_{i+1,j})/(\Delta x)^2 + (f_{i,j-1} - 2f_{i,j} + f_{i,j+1})/(\Delta x)^2 = -2 \tag{S29.7}$$

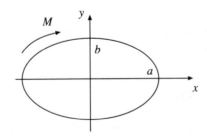

S13 Torsion of elliptic shaft

where use has been made of (S29.6). This relation, which applies to the internal points of the domain shown in Fig. S13, simplifies to

$$f_{i,j} = 1/4[f_{i-1,j} + f_{i+1,j} + f_{i,j-1} + f_{i,j+1} + 2(\Delta x)^2] \quad (S29.8)$$

In view of the symmetry about the x-axis, one gets from (S29.8) for the points $y = 0$

$$f_{0,j} = 1/4[2f_{1,j} + f_{0,j-1} + f_{0,j+1} + 2(\Delta x)^2] \quad (S29.9)$$

where $j = 1, 2, \ldots, n - 1$. Similarly, in view of the symmetry about the y-axis, one gets for the points $x = 0$

$$f_{i,0} = 1/4[f_{i-1,0} + f_{i+1,0} + 2f_{i,1} + 2(\Delta x)^2] \quad (S29.10)$$

where $i = 1, 2, \ldots, m - 1$, and for the centre ($j = 0$, $i = 0$)

$$f_{0,0} = 1/2[f_{1,0} + f_{0,1} + (\Delta x)^2] \quad (S29.11)$$

It remains to consider the boundary. The f-values taken on the points of the boundary must vanish to satisfy (S29.2). In the program given in the next section we use the interpolation of degree zero to incorporate this condition (see Section S28). A useful quantity for programming this part of the computations is denoted as $j_{\max}[i]$ and is the column subscript of the rightmost grid point in each row i, $i = 0, 1, \ldots, m$ (see Fig. S14). Since the equation of the ellipse is

$$1 - x^2/a^2 - y^2/b^2 = 0 \quad (S29.12)$$

$j_{\max}[i]$ is given by

$$j_{\max}[i] = n(1 - i^2/m^2)^{1/2} \quad (S29.13)$$

which is truncated to the next lower integer.

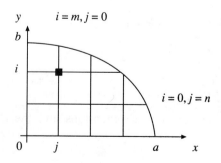

S14 Treating a quadrant of ellipse

S30. Torsion of a shaft, 2

(MAPLE file vfem69)

Turning to programming, state the relevant numerical values

> a:=10:b:=8:H:=1/3:M:=b/H:N:=a/H:

where M = *m* and N = *n* and then the equations for the points of the grid (see Fig. S14 of Section S29). First, equation (S29.9) for the points on the *x*-axis

> for j to N-1 do
> c[j]:=-f[0,j]+(2*f[1,j]+f[0,j-1]+f[0,j+1]+2*H^2)/4 od:

Also state the sequence of the associated unknowns and the equation for the corner point $i = 0, j = N$,

> f0y:=f[0,tt] $tt=1..N-1:
> f[0,N]:=0:

Repeat the same for the points on the *y*-axis using (S29.10) and then for corner point $i = M, j = 0$.

> for i to M-1 do
> q[i]:=-f[i,0]+(2*f[i,1]+f[i-1,0]+f[i+1,0]+2*H^2)/4 od:
> f0x:=f[xx,0] $xx=1..M-1:
> f[M,0]:=0:

State (S29.11) for the centre of the ellipse and then the associated unknown nodal value

> g[0]:=-f[0,0]+(f[1,0]+f[0,1]+H^2)/2:
> f00:=f[0,0]:

It remains to treat the interior and boundary points. To this end, formulate $j_{max}[i]$ as given by (S29.13)

> for i from 0 to M do
> jmax[i]:=floor(N*(1-i^2/M^2)^(1/2)) od:

and the equation (S29.8) for the interior points

> for i to M-1 do
> for j to jmax[i]-1 do
> r[i,j]:=-f[i, j]+(f[i-1,j]+f[i+1,j]+f[i,j-1]
  +f[i,j+1]+2*H^2)/4:od:od:

It is convenient to represent these equations and the associated unknown values as sequences

> for i to M-1 do
> k[i]:=r[i,d] $d=1..jmax[i]-1 od:

```
> for i to M-1 do
> ff[i]:=f[i,dd] $dd=1..jmax[i]-1 od:
> fint:=ff[x] $x=1..M-1:
```

For the points on and above the boundary the *f*-values must vanish. Therefore, state

```
> for i to M do p:= jmax[i]:
> f[i,p]:=0 od:
> for i to M do for j to N do if j>jmax[i] then f[i,j]:=0 fi:od:od:
```

completing thereby the formulation of the relevant equations.

Now set all the above equations as a sequence and check their number

```
> eqs:={c[jj] $jj=1..N-1,q[ii] $ii=1..M-1,g[0],
  k[zz]$zz=1..M-1}:
> nops(eqs);
    566
```

and then do the same for the unknown values

```
> unknowns:={f0y,f0x,f00,fint}:
> nops(unknowns);
    566
```

Solving this considerable system of equations, which may require a substantial computational time, one finds the *f*-values for the grid points,

```
> s:=evalf(solve(eqs,unknowns)):
> assign(s):
```

For example, for the points on the *x*-axis take the following values:

```
> f[0,hh] $hh=0..N;
```

38.51195788, 38.46700812, 38.33216132, 38.10742489, 37.79281091, 37.38833566, 36.89401901, 36.30988344, 35.63595288, 34.87225122, 34.01880066, 33.07561990, 32.04272217, 30.92011309, 29.70778820, 28.40572938, 27.01389916, 25.53223035, 23.96060707, 22.29882984, 20.54655351, 18.70318106, 16.76769053, 14.73837445, 12.61251074, 10.38618277, 8.055327948, 5.622831062, 3.133412914, .8389087840, 0

Resorting, for the purpose of comparison, to the exact solution given by (S29.5), one gets

```
> for j from 0 to N do x[j]:=H*j:
> y[j]:=0:F[j]:=(1-x[j]^2/a^2-y[j]^2/b^2)*(a^2*b^2/
  (a^2+b^2)) od:
> evalf(F[PP] $PP=0..N);
```

39.02439024, 38.98102981, 38.85094851, 38.63414634, 38.33062331, 37.94037940, 37.46341463, 36.89972900, 36.24932249, 35.51219512, 34.68834688, 33.77777778, 32.78048780, 31.69647696, 30.52574526, 29.26829268, 27.92411924, 26.49322493, 24.97560976, 23.37127371, 21.68021680, 19.90243902, 18.03794038, 16.08672087, 14.04878049, 11.92411924, 9.712737127, 7.414634146, 5.029810298, 2.558265583, 0

Comparison of the exact and approximate solutions shows a reasonable agreement, except, as may be anticipated, for the vicinity of the boundary. Using a finer grid or the interpolation of degree two (see Section S28), one may get more accurate results.

S31. Concluding remarks

Together with the variational and finite element methods, the finite difference technique constitutes the very core of engineering analysis. It is particularly convenient for treating transient phenomena studied in the theories of fluid flow, heat transfer, seismology, phase transformations, etc. In the case of nonlinear or chaotic systems, which are presently of special interest for researchers, this technique may provide the only calculational tool available for quantitative investigations. When combined with modern means of visualization, it yields the results in a form convenient for easy evaluation. This method may require the solution of a large system of simultaneous equations, which is one of the main subjects of numerical analysis and needs special treatment. Modern commercial codes are capable of treating systems of many thousands of equations.

W. Workshop

In this chapter we consider additional examples which illustrate the application of variational techniques.

In general, the presentation refers to results obtained via vfem-files with the extension .ms (MAPLE). To derive the relevant results with the help of MATHEMATICA the user must run the vfem-files with the extension .ma.

W1. Analysis of plates, 1

(MAPLE file vfem22)

Fig. W1 shows an elastic plate, which occupies the domain $0 \leq x \leq a$, $0 \leq y \leq d$ and is subjected to a transverse load q. The plate rests on an elastic foundation with the stiffness coefficient K^0.

The governing equation is

$$\nabla^4 w + Kw = w_{,xxxx} + 2w_{,xxyy} + w_{,yyyy} + Kw = q/D \qquad \text{(W1.1)}$$

where $w(x, y)$ is the deflection function, $K = K^0/D$ and $D = Eh^3/[12(1-v^2)]$ with h being the plate thickness and v Poisson's ratio.

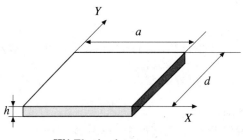

W1 Elastic plate

For the clamped plate no variations of $w(x,y)$ or its derivatives at the boundary are possible, as the all boundary conditions are specified. Consequently, the variation of the potential energy would contain no boundary terms, like those considered in Sections D3 and D23, and the problem may be solved, for example, by the conventional Bubnov–Galerkin method.

Seeking a second-order approximation, set

$$w(x,y) = v_1 w_1(x,y) + v_2 w_2(x,y) \tag{W1.2}$$

where v_1 and v_2 are adjustable coefficients and $w_1(x,y)$ and $w_2(x,y)$ must be so defined as to satisfy the boundary conditions. Substitute (W1.2) for $w(x,y)$ in (W1.1), multiply first by $w_1(x,y)$ and then by $w_2(x,y)$ and integrate over the plate in order to get the following two equations:

$$\begin{aligned} v_1 b_{11} + v_2 b_{12} &= c_1 \\ v_1 b_{21} + v_2 b_{22} &= c_2 \end{aligned} \tag{W1.3}$$

which can be also written as

$$v_i b_{ji} = c_j, \qquad i,j = 1,2 \tag{W1.4}$$

Here

$$b_{ji} = \int_0^a \int_0^d (\nabla^4 + K) w_j w_i \, dx \, dy \tag{W1.5}$$

$$c_j = \int_0^a \int_0^d (q/D) w_j \, dx \, dy \tag{W1.6}$$

The $b_{ji} = b_{ij}$, as (W1.5) shows. Solving (W1.3) for v_1 and v_2 and substituting back in (W1.2), one finds the second-order approximation for the deflection.

Going over to symbolic computation, state the following procedures defining the above b_{ji} and c_j:

```
> b:=proc(i,j,K)
> int(int((diff(w[i](x,y),x$4)+diff(w[i](x,y),y$4)
   +2*diff(w[i](x,y),x$2,
> y$2)+K*w[i](x,y))*w[j](x,y),x=0..a),y=0..d):
> end:
> c:=proc(j,q)
> int(int(q/sd*w[j](x,y),x=0..a),y=0..d):
> end:
```

where sd = D. It is seen that the procedures are so specified as to allow for variable K, q, i and j. On specifying these values and the trial functions $w_i(x,y)$, they provide the coefficients appearing in (W1.4).

The boundary conditions relevant for the clamped plate are

$$w([x=0, x=a], y) = w_{,x}([x=0, x=a], y) = 0$$
$$w(x, [y=0, y=d]) = w_{,y}(x, [y=0, y=d]) = 0 \qquad \text{(W1.7)}$$

which motivates the following coordinate functions:

$$w_1(x,y) = [xy(a-x)(d-y)]^2$$
$$w_2(x,y) = [xy(a-x)(d-y)]^2(x+y) \qquad \text{(W1.8)}$$

These series representations preserve the lowest powers possible from the viewpoint of the boundary conditions (W1.7).

Using (W1.3), consider the case $K = 0$ and $q = $ constant:

```
> f(x,y):=(x*y*(a-x)*(d-y))^2:
> w[1](x,y):=f(x,y):w[2](x,y):=f(x,y)*(x+y):
> eqs:={v1*b(1,1,0)+v2*b(1,2,0)=c(1,q),v1*b(1,2,0)
  +v2*b(2,2,0)=c(2,q)}:
> solve(eqs,{v1,v2});
```

$$\{v2 = 0,\ v1 = 49/8 \frac{q}{(7d^4 + 7a^4 + 4a^2d^2)\,sd}\}$$

It is seen that the second coordinate function, $w_2(x,y)$, contributes nothing to the solution ($v_2 = 0$). Indeed, because of the symmetry, the odd powers of x and y should disappear.

By situating the origin of the reference frame at the centre of the plate and assuming that this time its size is $2a \times 2d$ one may better incorporate the symmetry. Calculating the third-order approximation, set

$$w(x,y) = (x^2 - a^2)(y^2 - d^2)(v_1 + v_2 x^2 + v_3 y^2) \qquad \text{(W1.9)}$$

and turn to symbolic computation:

```
> f(x,y):=(x^2-a^2)^2*(y^2-d^{\,2})^2:w[1](x,y):=f(x,y):
> w[2](x,y):=f(x,y)*x^2:w[3](x,y):=f(x,y)*y^2:
```

State the equations (W1.4) for $i, j = 1, 2, 3$ and resolve them:

```
> eqs:={v1*b(1,1,0)+v2*b(1,2,0)+v3*b(1,3,0)=c(1,q),
> v1*b(1,2,0)+v2*b(2,2,0)+v3*b(2,3,0)=c(2,q),
> v1*b(1,3,0)+v2*b(2,3,0)+v3*b(3,3,0)=c(3,q)}:
> sol:=solve(eqs,{v1,v2,v3}):
```

This provides the deflection function $w(x,y)$:

```
> w:=f(x,y)*(v1+v2*x^2+v3*y^2):
> w:=subs(sol,w):
```

where the display of lengthy expressions has been prevented. In order to evaluate the accuracy, compute the deflection at the middle point of the square plate

```
> d:=a:x:=0:y:=0:
> w;
```

$$\frac{20713\, a^4\, q}{1025280\, sd}$$

```
> evalf(w);
```

$$.02020228620 \frac{a^4\, q}{sd} \qquad (\text{W}1.10)$$

which coincides with the exact solution up to three significant digits.

W2. Analysis of plates, 2

Consider a plate of arbitrary shape, which is subject to the lateral distributed load $q(x,y)$, boundary moments $m(c)$ and forces $Q(c)$, Fig. W2.

Then the potential of the external forces (including the boundary tractions) is

$$U_{\text{ext}} = -\iint_S q(x,y)w\,dx\,dy + \int_C [m(c)w_{,n} - Q(c)w]\,dc \qquad (\text{W}2.1)$$

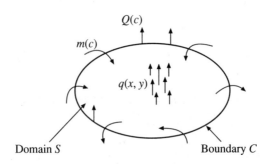

W2 To analysis of plate

where n is a unit normal to the boundary C. The strain energy is shown in the theory of plates to be

$$U_{\text{str}} = D/2 \iint_S [(w_{,xx} + w_{,yy})^2 + 2(1-v)(w_{,xy}^2 + w_{,xx}w_{,yy})]\,dx\,dy \quad (\text{W2.2})$$

Consequently, the potential energy to be extremized is given by

$$V = U_{\text{str}} + U_{\text{ext}} \quad (\text{W2.3})$$

As previously, set for the deflection $w(x,y)$ a trial function

$$w(x,y) = \sum_i A_i w_i(x,y) \quad (\text{W2.4})$$

Substituting (W2.4) in (W2.3) and integrating, we get

$$V = V(A_1, A_2, \ldots) \quad (\text{W2.5})$$

which yields the following condition of stationarity:

$$\partial V/\partial A_i = 0 \quad (\text{W2.6})$$

Complete the solution by finding A_i from the simultaneous system of equations given by (W2.6).

Consider again the clamped uniform plate (Fig. W1) under a uniform load q and set for the first approximation

$$w = A_1(1 - \cos 2\pi x/a)(1 - \cos 2\pi y/d) \quad (\text{W2.7})$$

which satisfies the relevant boundary conditions. Then (W2.1) and (W2.2) yield

$$U_{\text{str}} = DA_1^2 2\pi^4 (3a^4 + 3d^4 + 2a^2 d^2)/a^3 d^3 \quad (\text{W2.8})$$
$$U_{\text{ext}} = -A_1 q a d \quad (\text{W2.9})$$

Note that the contour integral in (W2.1) vanishes, as $w = w_{,n} = 0$ at the boundary. Equations (W2.3) and (W2.6) yield the value of A_1:

$$A_1 = qa^4 d^4 / [4D\pi^4 (3a^4 + 3d^4 + 2a^2 d^2)] \quad (\text{W2.10})$$

Further approximations stem from

$$w = \sum_i \sum_j A_{ij} [\cos 2i\pi x/a - \cos 2\pi(i+1)x/a]$$
$$[\cos 2j\pi y/d - \cos 2\pi(j+1)y/d] \quad (\text{W2.11})$$

Note that (W2.7) and (W2.10) yield for the deflection at the middle point of the square plate

$$w(x = a/2, y = a/2) = 1.28 \times 10^{-3} qa^4/D \quad (\text{W2.12})$$

which is larger than the exact result (see W1.10 where a denotes half the plate length). This confirms that the bounding property of the Rayleigh–Ritz method may not hold for the local value of the displacement. It can be shown that it is the work done by the external forces on the displacement found by this method, which remains less or equal to the exact value of the external work,

$$U_{\text{ext}}^{\text{Rayleigh}} \leq U_{\text{ext}}^{\text{exact}} \tag{W2.13}$$

This relation, in turn, leads to a possible over-estimation of the overall stiffness.

Note that only the essential (imposed) boundary conditions must be accounted for while applying the Rayleigh–Ritz method, which makes it particularly convenient for analysis of plates with a free edge, internal cuts, etc. Indeed, the boundary conditions in tractions are natural ones and their account stems from the stationarity of the relevant functional.

W3. Analysis of a shock-absorber, 1

The variational methods may be applied to problems of direct technological relevance. Consider the rubber-metal shock absorber shown in Fig. W3.

It is seen that it consists of a rubber block welded to the metal plates at the top and bottom. When subjected to force P, the shock-absorber height changes from $2h$ to $(2h - \Delta)$, which defines the overall stiffness K as

$$K = P/\Delta \tag{W3.1}$$

The reference system $x_1 x_2 x_3$ is placed at the centre of symmetry of the rubber block.

Experiments show that the linear theory of elasticity applies to rubber-like materials up to the deformations of about 10%. However, the incompressibility condition must be taken into account, particularly, for natural rubber. This implies that Poisson's ratio $\nu = 0.5$ and

$$\text{div } \mathbf{u} = u_{1,1} + u_{2,2} + u_{3,3} \tag{W3.2}$$

where $\mathbf{u}$ is the displacement vector function.

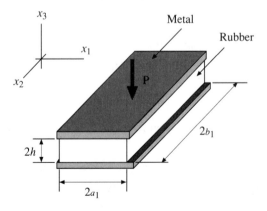

W3 Rubber-metal shock absorber

ANALYSIS OF A SHOCK-ABSORBER, 1   253

Turning to the problem at hand, note that the deformability of the metal plates may be neglected compared to that of the rubber and that the boundary conditions for the latter are of the mixed type. The external vertical planes of the rubber block are free of tractions, while at the interface with the metal plates the displacement must satisfy the following relations:

$$u_1(x_3 = h) = u_2(x_3 = h) = 0$$
$$u_3(x_3 = h) = -\Delta/2 \qquad \text{(W3.3)}$$

with the similar relations holding for $x_3 = -h$. Equation (W3.3) expresses a firm bonding at the interface between the metal and rubber.

The use of the Rayleigh–Ritz method seems particularly suitable, since the conditions in tractions, which are difficult to incorporate, may be neglected while constructing the trial functions. Indeed, these are the natural boundary conditions.

Since the plates are considered rigid, the strain energy is given by

$$U_{\text{str}} = \int_V (\lambda \theta^2/2 + \mu \epsilon_{ij} \epsilon_{ij}) \, dv \qquad \text{(W3.4)}$$

where V is the volume of the rubber block, λ and μ are the Lamé constants and

$$\theta = u_{1,1} + u_{2,2} + u_{3,3}$$
$$\epsilon_{ij} = (u_{i,j} + u_{j,i})/2 \qquad \text{(W3.5)}$$

Because of the incompressibility condition, $\theta = 0$ and the constant λ becomes unbounded. It is possible to show that, under these conditions, the first term in the integrand of (W3.4) vanishes and the strain energy simplifies to

$$U_{\text{str}} = \int_V \mu \epsilon_{ij} \epsilon_{ij} \, dv \qquad \text{(W3.6)}$$

where $\mu = G$ is the shear modulus. The work of the external forces is merely

$$U_{\text{ext}} = -P\Delta \qquad \text{(W3.7)}$$

Turning to the trial functions, note that the expressions for u_1 and u_2 may be taken as

$$u_1 = x_1(h^2 - x_3^2) \sum_i a_i x_3^i$$
$$u_2 = x_2(h^2 - x_3^2) \sum_i b_i x_3^i \qquad \text{(W3.8)}$$

so as to comply with the boundary conditions (W3.3). In turn, the expression for u_3 follows from the incompressibility condition

$$\theta = 0 \qquad \text{(W3.9)}$$

as

$$u_1 = -\int (u_{1,1} + u_{2,2})\,dx \qquad (W3.10)$$

Minimizing the governing functional by

$$\begin{aligned}\partial(U_{\text{str}} + U_{\text{ext}})/\partial a_i &= 0\\ \partial(U_{\text{str}} + U_{\text{ext}})/\partial b_i &= 0\end{aligned} \qquad (W3.11)$$

one may find an approximate solution to the problem.

W4. Analysis of a shock-absorber, 2

(MAPLE file vfem23)

Begin with a procedure which computes the density of the strain energy, namely the integrand of (W3.6). This quantity is denoted below as energy(u,v,w), where u, v and w are the displacement functions along the coordinate axes x_1, x_2 and x_3, respectively,

```
> energy:=proc(u,v,w)
> h[1]:=u:h[2]:=v:h[3]:=w:
> for i from 1 to 3 do
> for j from 1 to 3 do
> e[i,j]:=(diff(h[i],x[j])+diff(h[j],x[i]))/2:od:od:
> sum(sum(e[ii,jj]*e[ii,jj],ii=1..3),jj=1..3):
> end:
```

The first three statements in the above assign the values of u, v and w to the indexed variables $h[1]$, $h[2]$ and $h[3]$, respectively. Then the do-statement calculates the deformations $e[i,j] = e_{ij}$ according to (W3.5). Finally, the procedure sums their products. The integration of the value of this procedure over the volume of the rubber block would provide U_{str}.

In agreement with (W3.8) formulate u, v and w, which contain four free coefficients

```
> i:='i':j:='j':
> u:=x[1]*(h^2-x[3]^2)*sum(a[i]*x[3]^i,i=0..1);
```

$$u := x[1]\,(h^2 - x[3]^2)\,(a[0] + a[1]\,x[3])$$

```
> v:=x[2]*(h^2-x[3]^2)*sum(b[j]*x[3]^j,j=0..1);
```

$$v := x[2]\,(h^2 - x[3]^2)\,(b[0] + b[1]\,x[3])$$

```
> w:=-int(diff(u,x[1])+diff(v,x[2]),x[3]);
```

ANALYSIS OF A SHOCK-ABSORBER, 2   255

$$w := 1/4\ a[1]\ x[3]^4 + 1/3\ a[0]\ x[3]^3 - 1/2\ h^2\ a[1]\ x[3]^2 - h^2\ a[0]x[3]$$
$$+ 1/4\ b[1]\ x[3]^4 + 1/3\ b[0]\ x[3]^3 - 1/2\ h^2\ b[1]\ x[3]^2 - h^2\ b[0]x[3]$$

where use has been made of (W3.10) to state w. Now check the incompressibility condition

```
> simplify(diff(u,x[1])+diff(v,x[2])+diff(w,x[3]));
    0
```

and find the overall strain energy

```
> Ustr:=8*G*int(int(int(energy(u,v,w),x[1]=0..a1),
  x[2]=0..b1),x[3]=0..h):
```

The expression for Δ is

```
> delta:=simplify(-2*subs(x[3]=h,w));
```

$$delta := 1/2\ a[1]\ h^4 + 4/3\ a[0]\ h^3 + 1/2\ b[1]\ h^4 + 4/3\ b[0]\ h^3$$

and the functional of the potential energy F is

```
> F:=Ustr-P*delta:
```

Minimizing this value, we get the system of four simultaneous equations and resolve it:

```
> for k from 0 to 1 do r[k]:=diff(F,a[k]):t[k]:=diff(F,b[k]):
  od:
> d22:=solve({r[0],r[1],t[0],t[1]},{a[0],a[1],b[0],
  b[1]}):
> Digits:=3;
      Digits := 3
```

It remains to substitute this solution in the above expression for Δ to complete the investigation:

```
> subs(d22,delta):delta:=simplify(");
```

$$delta := 5/2\ P\ h^3\ (513216\ h^6 + 1536180\ h^4\ b1^2 + 3427256\ a1^2\ b1^2\ h^2$$
$$+ 301840\ b1^4\ h^2 + 588245\ a1^2\ b1^4 + 1536180\ a1^2\ h^4$$
$$+ 301840\ a1^4\ h^2 + 588245\ a1^4\ b1^2)/(G\ a1\ b1\ (26570592\ a1^2\ h^6$$
$$+ 2941225\ a1^4\ b1^4 + 65605368\ a1^2\ h^4\ b1^2 + 26570592\ h^6\ b1^2$$
$$+ 8503056\ h^8 + 6667920\ b1^4\ h^4 + 15627080\ a1^2\ b1^4\ h^2$$
$$+ 6667920\ a1^4\ h^4 + 15627080\ a1^4\ h^2\ b1^2))$$

It may be useful to derive a simpler result for the plane problem, which arises if the shock absorber is a long one, b1 $\to \infty$, by asympt

> asympt(delta,b1,2);

$$5/2\,\frac{P\,h^3\,(588245\,a1^2 + 301840\,h^2)}{G\,a1\,(2941225\,a1^4 + 6667920\,h^4 + 15627080\,a1^2\,h^2)\,b1} + O(\frac{1}{b1^3})$$

W5. Flow through a duct

(MAPLE file vfem24)

Fig. W4 shows a long square duct subjected to a flow of the Newtonian fluid. The velocity components are

$$v_1 = v_2 = 0, \qquad v_3 = v(x_1, x_2) \tag{W5.1}$$

and the governing equation is

$$v_{,11} + v_{,22} - p_{,3}/\mu = 0 \tag{W5.2}$$

Here p is the pressure and μ viscosity. Since the velocity vanishes at the duct walls, the boundary conditions are

$$v(x_1 = a, x_2) = v(x_1, x_2 = a) = 0 \tag{W5.3}$$

with the similar relations for $x_1 = -a$ and $x_2 = -a$. Note that the last term in (W5.2) is a constant, which may be conveniently denoted as

$$p_{,3}/\mu = m/2 \tag{W5.4}$$

The direct verification shows that (W5.2) is the Euler–Lagrange equation for the following functional:

$$Q = \int_{-a}^{a}\int_{-a}^{a} (v_{,1}^2 + v_{,2}^2 + mv)\,dx_1\,dx_2 \tag{W5.5}$$

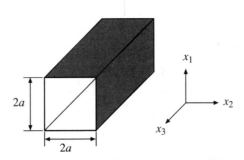

W4 Square long duct

where m is given by (W5.4). Therefore, one may seek for a stationary value of Q, instead of solving (W5.2).

A first approximation may be given by

$$v = (x_1^2 - a^2)(x_2^2 - a^2)b_0 \tag{W5.6}$$

Taking into account the symmetry, set a more general expression as

$$v = (x_1^2 - a^2)(x_2^2 - a^2)[b_0 + b_1(x_1^2 + x_2^2) + b_2 x_1^2 x_2^2] \tag{W5.7}$$

Substituting this for v in (W5.5) and then taking the derivatives of Q with respect to the free coefficients, get the system of three simultaneous equations, solution of which specifies the third approximation to the velocity field.

Firstly, state (W5.7) and then the functional Q, as given by (W5.5):

```
> v:=(x1^2-a^2)*(x2^2
  -a^2)*(b[0]+b[1]*(x1^2+x2^2)+b[2]*x1^2*x2^2);
```

$$v := (x1^2 - a^2)(x2^2 - a^2)(b[0] + b[1](x1^2 + x2^2) + b[2] x1^2 x2^2)$$

```
> diff(v,x1)^2+diff(v,x2)^2+m*v:
> Q:=int(int(",x1=-a..a),x2=-a..a);
```

Take the derivatives, resolve the system of equations and substitute the result in v to obtain the solution

```
> for i from 0 to 2 do k[i]:=diff(Q,b[i]) od:
> solve({k[0],k[1],k[2]},{b[0],b[1],b[2]}):
> subs(",v);
```

$$(x1**2-a**2)*(x2**2-a**2)*(-151/1024/a**2*m-21/1024/a**4*m*(x1**2+x2**2)-63/1024/a**6*m*x1**2*x2**2)$$

Compute also the average velocity given by

$$v_{av} = \int_{-a}^{a} \int_{-a}^{a} v \, dx_1 \, dx_2 / (4a^2) \tag{W5.8}$$

or, in the MAPLE language,

```
> int(int(",x1=-a..a),x2=-a..a)/(4*a^2):evalf(");
```

$$-.07027777778 \, m \, a^2$$

In order to evaluate the accuracy note that, say, for $x_1 = x_2 = 0.4a$ the exact value of (v/v_{av}) is 1.561, while the present solution provides this value as 1.562.

W6. Free vibrations by the Rayleigh–Ritz method

Applications of direct methods to dynamic problems may be particularly useful in view of their complexity. We overview below the main aspects of an approximate analysis of free vibrations.

In this case, a generalized coordinate η_i of the conservative system may obey the law

$$\eta_i = u_i \sin \omega_n t \tag{W6.1}$$

where u_i is the amplitude and ω_n the natural frequency, both of which are unknown. Then the kinetic energy T is

$$T = a_{ij}\eta_{i,t}\eta_{j,t} \tag{W6.2}$$

where a_{ij} are coefficients depending on the system and summation with respect to the repeated subscript applies. Substituting (W6.1) in (W6.2), we get

$$T = \omega_n^2 m_0 \cos^2(\omega_n t)/2 \tag{W6.3}$$

where the generalized mass m_0 is

$$m_0 = a_{ij} u_i u_j \tag{W6.4}$$

The potential energy U is given by

$$U = U_{\max} \sin^2 \omega_n t \tag{W6.5}$$

with $U_{\max}$ being its maximal value.

Hamilton's principle (see Section D2) states

$$\delta \int_{t_1}^{t_2} (T - U)\, dt = 0 \tag{W6.6}$$

Since the time instants t_1 and t_2 do not vary, we may consider the time interval $(t_2 - t_1)$ to be the natural period of vibrations $T_n = 2\pi/\omega_n$. Substituting (W6.3) and (W6.5) in (W6.6) and integrating from $t_1 = 0$ to $t_2 = T_n$, we get

$$\delta(\omega_n^2 m_0/2 - U_{\max}) = 0 \tag{W6.7}$$

In the context of direct methods, these results apply, with slight generalizations, to continuous systems too. To this end, represent the 'amplitude' u_i as a conventional series

$$u_i = a_1 u_{i1} + a_2 u_{i2} + \cdots \tag{W6.8}$$

where spatial functions u_{ij} satisfy the boundary conditions. Then the stationarity condition (W6.6) delivers the linear homogeneous system of simultaneous equations

$$(\omega_n^2/2)\, \partial m_0/\partial a_i - \partial U_{\max}/\partial a_i = 0 \qquad i = 1, 2, \ldots \tag{W6.9}$$

FREE VIBRATIONS BY THE RAYLEIGH–RITZ METHOD

The condition of vanishing determinant $f(\omega_n^2) = 0$ provides the frequency equation for ω_n. The number of roots, in general, corresponds to the number of degrees of freedom, which in turn is governed by the number of free coefficients in (W6.8). The substitution of these roots back into (W6.9) leads to identification of the modes of vibrations.

In the case of the first approximation, this procedure simplifies to

$$\omega_n^2 = U_{\max}/T^0 \tag{W6.10}$$

with $T^0 = T_{\max}/\omega_n^2$. Indeed, the first term of the bracketed expression in (W6.7) is the maximal kinetic energy $T_{\max}$.

As noted earlier, the Rayleigh–Ritz method provides the upper bound for the potential energy of the system. The above result (W6.10) suggests that the same is true for ω_n. Equation (W6.9) now appears as a means of minimizing this bound.

W7. Applications, 1

As an application, consider a hinged beam. The boundary conditions for the deflection function w are as follows:

$$w(x = 0) = w(x = L) = w_{,xx}(x = 0) = w_{,xx}(x = L) = 0. \tag{W7.1}$$

In the spirit of (W6.1) and (W6.8), set

$$w = u \sin \omega_n t \tag{W7.2}$$

with

$$u = a_1 \sin \pi x/L + a_2 \sin 2\pi x/L \tag{W7.3}$$

The last expression implies that the beam is approximated as a system with two degrees of freedom, with a_1 and a_2 being the generalized coordinates.

Generalizing (W6.4), set

$$m = \int_0^L \rho u^2 \, dx \tag{W7.4}$$

with ρ being the mass density per unit length. Equation (W7.2) yields for the maximum of strain energy

$$U_{\max} = \int_0^L EI u_{,xx}^2 \, dx/2 \tag{W7.5}$$

Consider the case of variable ρ and EI:

$$\rho = \rho_0[1 - x/(2L)], \qquad EI = J_0[1 - x/(2L)] \tag{W7.6}$$

WORKSHOP

With these expressions, (W7.4) and (W7.5) provide

$$m_0 = \rho_0 L[a_1^2/8 + 8a_1a_2/(9\pi^2) + 3a_2^2/8] \tag{W7.7}$$

$$U_{max} = \pi^4 J_0/(2L^3)[3a_1^2/8 + 32a_1a_2/(9\pi^2) + 6a_2^2] \tag{W7.8}$$

Substituting these in (W6.9), we get the system of two homogeneous equations with the determinant

$$9(\beta^2 - 1)(\beta^2 - 16)/16 - 64(\beta^2 - 4)^2/(81\pi^4) \tag{W7.9}$$

where

$$\beta = \omega_n^2 \rho_0 L^4/(\pi^4 J_0) \tag{W7.10}$$

This determinant vanishes for

$$\beta^2 = 0.9913, \qquad \beta^2 = 16.14 \tag{W7.11}$$

(check this by solving (W7.9) with the help of MAPLE), which specifies the two natural frequencies of vibrations.

W8. Applications, 2

(MAPLE file vfem26)

The analysis given in Sections W6 and W7 may be performed in a generalized and automated way with the help of MAPLE. The master program below is similar to that of Section D7.

Specify first a polynomial for the deflection w, then its first and second derivatives and the expressions involved in the boundary conditions,

```
> d1:=sum(a[j]*x^j,j=1..5):
> d2:=diff(",x):
> d3:=diff(",x):
> fL:=subs(x=L,d1):f10:=subs(x=0,d2):f1L:=subs(x=L,d2):
> f20:=subs(x=0,d3):f2L:=subs(x=L,d3):
```

Further, load the linalg package, denote wn2 = ω_n^2, m0 = m_0, Umax = U_{max}, r = ρ and state, say, the rid procedure which embodies all the operations needed to find the natural frequencies, as given in the above considerations, and (W7.4), (W7.5) and (W6.9). Note that the rid procedure has, as its arguments, the 'amplitude' u, density ρ and stiffness EI:

```
> with(linalg,genmatrix,det):
    #wn2 = wn^2.
> rid:=proc(u,r,EI)
> diff(u,x,x):Umax:=int("^2/
    2*EI,x=0..L):m0:=int(u^2*r,x=0..L):
```

```
> for i from 4 to 5 do b[i]:=wn2*diff(m0,a[i])/2
    -diff(Umax,a[i]) od:
> genmatrix({b[4],b[5]},{a[4],a[5]}):det("):solve(",wn2);
> end:
```

Try first the problem considered in Section W7. Since the deflection w as given by the above d1 satisfies the condition $w(x = 0) = 0$, treat the rest of the boundary conditions (W7.1) and then specify the trial function for w:

```
> d7:=solve({fL,f20,f2L},{a[1],a[2],a[3]}):
> w:=subs(",d1);
w :=
```

$$(7/3\ a[5]\ L^4 + a[4]\ L^3)\ x - 2/3L\ (3\ a[4] + 5\ a[5]\ L)\ x^3 + a[4]\ x^4 + a[5]\ x^5$$

It is seen that w contains two free coefficients and the system is thus of two degrees of freedom. In order to obtain the natural frequences invoke the rid procedure with the properly specified arguments, and apply evalf:

```
> f:=1-x/(2*L):
> d9:=rid(w,r0*f,J0*f):map(evalf,op("));
```

$$\text{op}(1600.\ \frac{J0}{r0\ L^4},\ 97.6\ \frac{J0}{r0\ L^4})$$

These results are in full agreement with those of Section W7, a slight difference being caused by the use of a polynomial for w instead of trigonometric functions.

For the clamped uniform beam we get

```
> d15:=solve({fL,f10,f1L},{a[1],a[2],a[3]}):
> w:=subs(",d1):
> rid(w,r,EI);
```

$$3960\ \frac{EI}{r\ L^4},\ 504\ \frac{EI}{r\ L^4}$$

For the exact solution the above coefficients are 3806.89 and 501.76, respectively. Note, the Rayleigh–Ritz estimates are indeed higher than the exact results.

This program easily modifies to the case of a beam resting on an elastic foundation. Then (W7.5) takes the form

$$U_{\max} = \int_0^L (EIu_{xx}^2 + Ku^2)\,dx/2 \qquad (W8.1)$$

with K being the stiffness of the foundation. A proper ridel procedure, which may be produced by editing the above rid procedure, is given below:

```
> ridel:=proc(u,r,EI,K)
> diff(u,x,x):Umax:=int("^2/2*EI
  +u^2/2*K,x=0..L):m0:=int(u^2*r,x=0..L):
> for i from 4 to 5 do b[i]:=wn2*diff(m0,a[i])/2-diff(Umax,
  a[i]) od:
> genmatrix({b[4],b[5]},{a[4],a[5]}):det("):solve(",wn2);
> end:
```

This program computes the natural frequencies in an automated way provided the arguments are given. Exercises are left to the reader.

W9. Free vibrations by the Bubnov–Galerkin method

(*MAPLE file vfem27*)

Consider longitudinal free vibrations of a non-uniform bar, for which the density and cross-sectional area vary as

$$\rho = \rho_0(1 + x/L), \qquad A = A_0(1 + x/L) \tag{W9.1}$$

with L being the bar length. Denoting the axial displacement function as $u_a(x, t)$, set the governing equation as

$$d(EAu_{a,x})/dx - \rho u_{a,tt} = 0 \tag{W9.2}$$

where the first term is due to the elastic force and the second to the inertial force. The boundary conditions are

$$u_a(x = L) = 0, \qquad u_{a,x}(x = 0) = 0 \tag{W9.3}$$

which describe the free end at $x = 0$ and the clamped end at $x = L$.

Using the separation of variables, set

$$u_a(x, t) = f(x) \sin \omega_n t \tag{W9.4}$$

and substitute this into the governing equation (W9.2) to obtain

$$d(EAf_{,x})/dx + \rho \omega_n^2 f = 0 \tag{W9.5}$$

Expand $f(x)$ as

$$f(x) = a_1 f_1(x) + a_2 f_2(x) + \cdots + a_k f_k(x) \tag{W9.6}$$

where $f_i(x)$ must be so chosen as to meet the boundary conditions (W9.3), and substitute this into (W9.5) to arrive at the residual, say, $F(x) \neq 0$. The Bubnov–Galerkin method delivers the system of simultaneous homogeneous equations

$$\int_0^L F(x) f_i(x)\, dx = 0, \qquad i = 1, 2, \ldots, k \tag{W9.7}$$

the vanishing determinant of which yields, in turn, the desirable frequency equation.

FREE VIBRATIONS BY THE BUBNOV–GALERKIN METHOD

Confining the investigation to the first two frequencies, set

$$f_1(x) = (1 - x^2/L^2), \qquad f_2(x) = (1 - x^3/L^3) \tag{W9.8}$$

which implies that the displacement (W9.4) saisfies the boundary conditions (W9.3). Equations (W9.7) and (W9.5) provide two simultaneous equations:

$$\int_0^L [d(EAf_{i,x})/dx + \rho\omega_n^2 f_i]f_j\,dx = 0, \quad i,j = 1,2 \tag{W9.9}$$

Denoting

$$\int_0^L \rho f_i f_j\,dx = T_{ij}$$
$$\int_0^L [d(EAf_{i,x})/dx]f_j\,dx = V_{ij} \tag{W9.10}$$

reduce the determinantal equation to the form

$$(V_{11} + \omega_n^2 T_{11})(V_{22} + \omega_n^2 T_{22}) - (V_{12} + \omega_n^2 T_{12})^2 = 0 \tag{W9.11}$$

Turning to symbolic computations, state $f_i(x)$, T_{ij} and V_{ij}

```
> y:=1+x/L;
    y:= 1 + x/L
> T := array (1..2,1..2): V := array(1..2,1..2):
> for i from 1 to 2 do
> f[i](x):=1-(x/L)^(i+1);
> for j from
> 1 to 2 do
> f[j](x):=1-(x/L)^(j+1);
> T[i,j]:=int(r0*y*f[i](x)*f[j](x),x=0..L):
> diff(f[j](x),x):p[j](x):=diff(e*A0*y*",x):
> V[i,j]:=int(p[i](x)*f[j](x),x=0..L) od od:
```

with e = E, ro = ρ, A0 = A_0

It remains to formulate the determinant and solve for the natural frequencies (wn2 = ω_n^2)

```
> with(linalg,det,add):
> add(T,V,wn2,1);det("):
```

$$\begin{bmatrix} 7/10 \ wn2 \ r0 \ L - 7/3 \dfrac{e \ A0}{L} & \dfrac{163}{210} \ wn2 \ r0 \ L - \dfrac{27}{10} \dfrac{e \ A0}{L} \\ \dfrac{163}{210} \ wn2 \ r0 \ L - \dfrac{27}{10} \dfrac{e \ A0}{L} & \dfrac{243}{280} \ wn2 \ r0 \ L - \dfrac{33}{10} \dfrac{e \ A0}{L} \end{bmatrix}$$

```
> s:=solve(",wn2);
```

$$s := 1/1774 \ \dfrac{25326 \ r0 \ L^2 \ e \ A0 + 42 \ 218141^{1/2} \ r0 \ L^2 \ e \ A0}{r0^2 \ L^4}$$

$$1/1774 \ \dfrac{25326 \ r0 \ L^2 \ e \ A0 - 42 \ 218141^{1/2} \ r0 \ L^2 \ e \ A0}{r0^2 \ L^4}$$

```
> evalf(s[1]);evalf(s[2]);
```

$$25.33389977 \ \dfrac{e \ A0}{r0 \ L^2}$$

$$3.218524132 \ \dfrac{e \ A0}{r0 \ L^2}$$

On the other hand, the first approximation would yield

$$V_{11} + \omega_n^2 T_{11} = 0 \qquad (W9.12)$$

with the natural frequency $\omega_n^2 = 3.3333 E A_0/\rho_0$, which shows that the second approximation not only delivers the second frequency, but also improves the estimation for the first frequency.

W10. Non-linear free vibrations

(MAPLE file vfem28)

Duffing's equation

$$m x_{,tt} + k x + k_1 x^3 = 0 \qquad (W10.1)$$

is a generalization of the classical equation of harmonic oscillations

$$m x_{,tt} + k x = 0 \qquad (W10.2)$$

which holds for small x. Equation (W10.1) describes the free vibrations of mass

m under the elastic restoring force $F(x)$ given by

$$F(x) = kx + k_1 x^3 \tag{W10.3}$$

Looking for a periodic solution, set

$$x = A\cos\omega t \tag{W10.4}$$

which is an exact solution to (W9.2) provided $\omega = (k/m)^{1/2} = 2\pi/T$ with T being the period of free vibrations.

Denoting

$$f(x) = F(x)/m, \qquad \omega_n^2 = k/m \tag{W10.5}$$

rewrite (W10.1) as

$$x_{,tt} + \omega_n^2(x + m_1 x^3) = 0 \tag{W10.6}$$

with

$$m_1 = k_1/(m\omega_n^2) \tag{W10.7}$$

Considering (W10.4) a trial function, substitute it in (W10.6), multiply the result by $\cos\omega t$, integrate over the period T and solve for ω^2, all this in the spirit of the Bubnov–Galerkin method.

In the symbolic computations below $w = \omega$, $w2 = \omega^2$, $wn2 = \omega_n^2$ and $m1 = m_1$

> d2:=diff(x(t),t,t)+wn2*(x(t)+m1*x(t)^3);

$$d2 := \left(\frac{d^2}{dt^2} x(t)\right) + wn2\,(x(t) + m1\,x(t)^3)$$

> d3:=subs(x(t)=A*cos(w*t),d2);

$$d3 := \left(\frac{d^2}{dt^2} A\cos(w\,t)\right) + wn2\,(A\cos(w\,t) + m1\,A^3 \cos(w\,t)^3)$$

> d4:=int(d3*cos(w*t),t=0..2*Pi/w)=0;w2:=solve(",w^2);

$$d4 := 1/4\,\frac{A\,\text{Pi}\,(4\,wn2 - 4\,w^2 + 3\,wn2\,m1\,A^2)}{w} = 0$$

$$w2 := wn2 + 3/4\,wn2\,m1\,A^2$$

The last expression

$$\omega^2 = \omega_n^2 + 3\omega_n^2 m_1 A^2/4 \tag{W10.8}$$

shows a dependence between the amplitude A and frequency ω. For small amplitudes this trivially yields $\omega^2 = \omega_n^2$.

W11. Applications of Gauss' principle

(*MAPLE file vfem29*)

Section D14 deals with a description of the weighted residual approach from a fairly formal viewpoint. One of the ways of arriving at the physically meaningful form of the weighted residual is the Bubnov–Galerkin method. Yet another way is application of Gauss' principle (see Section D25).

Consider again free non-linear vibrations governed by

$$x_{,tt} + \omega_n^2 x + \epsilon x^3 = 0, \quad \epsilon \ll 1 \tag{W11.1}$$

According to Gauss' principle, the unknown function $x(t)$ would provide the minimum to the constraint (residual) Z given by

$$Z = (x_{,tt} + \omega_n^2 x + \epsilon x^3)^2 \tag{W11.2}$$

on the understanding that only the acceleration $x_{,tt}$ is involved in the minimization process. This means that if, say, one looks for a solution in the form

$$x(t) = A \cos \omega t \tag{W11.3}$$

the above Z-constraint must be minimized with respect to the acceleration terms only, unlike the method of least squares, which considers all the terms equally important. Further, as Z may depend on time, (W11.2) might be replaced by the integral expression

$$Z_0 = \int_0^{2\pi/\omega} Z \, dt \tag{W11.4}$$

where $2\pi/\omega$ is the period of vibrations.

Technically it may be convenient to substitute (W11.3) in the 'force' term

$$\omega_n^2 x + \epsilon x^3 \tag{W11.5}$$

only, while substituting, say,

$$x(t) = B \cos \omega t \tag{W11.6}$$

in the 'acceleration' term $x_{,tt}$. Then the resolving equation is

$$dZ_0/dB = 0 \tag{W11.7}$$

Recovering $B = A$ *after* the minimization process and solving for ω, one completes the solution.

Going over to symbolic computation, state first the constraint Z in the notations w = ω, wn2 = ω_n and e = ϵ,

```
> x:=A*cos(w*t);
> accel:=diff(x,t,t);
```

APPLICATIONS OF GAUSS' PRINCIPLE   267

$$\text{accel} := -A\cos(wt)w^2$$

```
> eq:=subs(A=B,")+wn2*x+e*x^3;
```

$$eq := -B\cos(wt)w^2 + wn2\,A\cos(wt) + eA\cos(wt)^3$$

```
> Z:=eq^2:
```

The quantity Z contains two coefficients A and B. Next, formulate Z_0 and differentiate with respect to B:

```
> Z0:=int(eq^2,t=0..2*Pi/w);
> diff(Z0,B);
```

$$\frac{1}{8}\frac{\text{Pi}\,(-16w^2\,wn2\,A + 16\,Bw^4 - 12w^2eA^3)}{w}$$

Recover $B = A$ and solve for $w2 = \omega^2$:

```
> eq1:=subs({B=A,w=sqrt(w2)},");
```

$$eq1 := \frac{1}{8}\frac{\text{Pi}\,(-16\,w2\,wn2\,A + 16\,A\,w2^2 - 12\,w2\,eA^3)}{w2^{1/2}}$$

```
> solve(",w2);
```

$$0,\ wn2 + 3/4\,eA^2,\ wn2 + 3/4\,eA^2$$

which may be conveniently represented as a list:

```
> ["]:evalf(");
```

$$[0,\ wn2 + .7500000000\,eA^2,\ wn2 + .7500000000\,eA^2]$$

This coincides with the earlier result (W10.8).

For comparison solve the same problem by the least squares method, in which case all the terms participate in the minimization process,

```
> eq:=diff(x,t,t)+wn2*x+e*x^3;
```

$$eq := -A\cos(wt)w^2 + wn2\,A\cos(wt) + eA^3\cos(wt)^3$$

```
> res:=int(eq^2,t=0..2*Pi/w):
> diff(res,A):eq2:=normal("):
> subs(w=sqrt(w2),eq2):
> solve(",w2):
> ["]:evalf(");
```

$$[wn2 + 2.112372436\,eA^2,\ wn2 + 2.112372436\,eA^2,\ wn2 + .8876275642\,eA^2,$$
$$wn2 + .8876275642\,eA^2]$$

where the lower frequency is physically meaningful.

Thus, unlike (W10.8) and the above prediction of Gauss' method, the least squares technique delivers a crude result

$$\omega^2 = \omega_n^2 + 0.89\epsilon A^2 \tag{W11.8}$$

W12. Heat transfer to fluid

(MAPLE file vfem30)

Though from a purely formal point of view Gauss' principle deals with acceleration, under a proper interpretation it may apply to a broader class of problems. As an example, consider heat (mass) transfer to a fluid in stagnation flow toward a flat interface (Fig. W5). The initial and boundary conditions for the temperature T may be set as follows:

$$\begin{array}{lll} T = 0 & \text{at} & t = 0 \\ T = 1 & \text{at} & x_2 = 0 \\ T \to 0 & \text{at} & x_2 \to \infty \end{array} \tag{W12.1}$$

and the governing equation as follows:

$$T_{,\theta} + 2P_e x T_{,x} - 2P_e y T_{,y} = T_{,xx} + T_{,yy} \tag{W12.2}$$

Here $\theta = t\kappa/L^2$ is the dimensionless time, $x = x_1/L$ and $y = x_2/L$ are the dimensionless coordinates, P_e is the Peclet number and κ is diffusivity with L being a characteristic length. Introducing a boundary layer of thickness q, set

$$T = 1 + \sum_{i=1}^{N} a_i(y/q)^i \quad \text{for} \quad 0 \leq y \leq q \tag{W12.3}$$

$$T = 0 \quad \text{for} \quad y \geq q$$

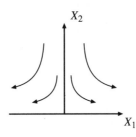

W5 Fluid in stagnation flow

Matching these two expressions at $y = q$, we get

$$0 = 1 + \sum_{i=1}^{N} a_i \qquad (W12.4)$$

The 'smoothness' at the $y = q$ implies

$$T_{,y} = 0 \quad \text{at} \quad y = q \qquad (W12.5)$$

Equations (W12.4) and (W12.5) suggest the following simple form for the temperature T:

$$T = 1 - 2y/q + (y/q)^2 \qquad (W12.6)$$

Note, in the above the layer thickness q is a function of time, $q = q(t)$.

In view of (W12.6), we get from (W12.2)

$$T_{,\theta} - 2P_e y T_{,y} = T_{,yy} \qquad (W12.7)$$

Regardless of the specific physical content of this expression, one may consider, say, its right-hand side as the 'acceleration' term and its left-hand side as the 'force' term. Thus, Gauss' constraint Z_0 is

$$Z_0 = \int_0^q (T_{,\theta} - 2P_e y T_{,y} - T_{,yy})^2 \, dy \qquad (W12.8)$$

on the understanding that only the 'acceleration' term $T_{,yy}$ is subject to variation.

Below we state (W12.6) and the integrand of (W12.8):

```
> T:=1-2*y/q(t)+(y/q(t))^2;
```

$$T := 1 - 2\frac{y}{q(t)} + \frac{y^2}{q(t)^2}$$

```
> s:=K=diff(T,y,y)/2;
```

$$s := K = \frac{1}{q(t)^2}$$

```
> (diff(T,t)-2*Pe*y*diff(T,y)-2*K)^2:
```

where $K = T_{,yy}/2$. Thus, Z_0 should be minimized with respect to K:

$$Z_{0,K} = 0 \qquad (W12.9)$$

Facilitating the further programming, make the substitutions $q1 = q_{,t}$ and $q = q(t)$ in the above integrand, then find Z_0 in agreement with (W12.8):

```
> subs([diff(q(t),t)=q1,q(t)=q],");
> Z0:=int(",y=0..q);
```

$$Z0 := 2/15 \frac{-10q1Kq + 30K^2q^2 + q1^2 + 4Pe^2q^2 - 20Peq^2K + 4q1Peq}{q}$$

and take the derivative (W12.9):

> diff(Z0,K);

$$2/15 \frac{-10q1q + 60Kq - 20Peq^2}{q}$$

At this stage recover the original notations:

> subs(K=1/q^2,");

$$2/15 \frac{-10q1q + 60 - 20Peq^2}{q}$$

> eq:=subs({q=q(t),q1=diff(q(t),t)},");

$$eq := 2/15 \frac{-10\left(\dfrac{d}{dt}q(t)\right)q(t) + 60 - 20Peq(t)^2}{q(t)}$$

and solve this equation subject to the initial condition $q(t=0) = 0$:

> sol:=dsolve({eq=0,q(0)=0},q(t));
sol :=

$$q(t) = -1/2 \frac{(12 - 12\exp(-4Pet))^{1/2}}{Pe^{1/2}},$$

$$q(t) = 1/2 \frac{(12 - 12\exp(-4Pet))^{1/2}}{Pe^{1/2}}$$

Choosing the positive solution, set

$$q(t) = \{3[1 - \exp(-4P_e\theta)]/P_e\}^{1/2} \qquad \text{(W12.10)}$$

which, with the help of (W12.6), delivers the Nusselt number Nu

$$Nu = -T_y(y = 0) = 2/q(t) \qquad \text{(W12.11)}$$

The comparison with the exact result shows that the above solution is of satisfactory accuracy. For example, the exact Nusselt number is $Nu_{\text{exact}} = Nu/1.02$. Note that from a formal viewpoint one may consider the left-hand side of (W12.7) as the 'acceleration' term and its right-hand side as the 'force' term, which may lead to a different solution. Generally, it is difficult to assess which one of the two versions of this approach provides a better accuracy.

Problems

Chapter D

1. Find the variation of the functional I given by

$$I = \int_0^L w_{,xx}^2 \, dx$$

2. Consider a clamped beam under a distributed load and derive the natural boundary conditions.

3. Consider a simply supported beam and derive the natural boundary conditions.

4. Heat conduction in a two-dimensional medium may be described by the functional

$$F = \iint_S [(T_{,x}^2 + T_{,y}^2) - QT] \, dx \, dy$$

where T is the temperature and Q the rate of heat generation. Derive the Euler–Lagrange equation.

5. Derive the Euler–Lagrange equation for the functional

$$I = \int_0^L (y_{,x}^2 - y^2) \, dx$$

6. Derive the stationarity condition for the functional

$$I = \int_0^L (1 - y_{,x}^2)^{1/2} / y^{1/2} \, dx$$

7. Investigate a cantilever by the Bubnov–Galerkin method (first approximation) and compare the result with the exact solution.

8. Investigate a cantilever by the Rayleigh–Ritz method (first approximation) and compare the result with the previous one.

9. With the help of the master program (Section D7) obtain a solution for: (a) non-uniform cantilever, (b) clamped non-uniform beam, (c) clamped-hinged non-uniform beam. The stiffness $EI = z + gx$ and the load $q = sx^2$, where z, g and s are constants. Solve the same problems with the help of master program given in the end of Section D8.

10. Derive a solution by the Rayleigh–Ritz method (first approximation) for the shock-absorber considered in Sections W3 and W4, assuming from the beginning that the displacement along the x_2-axis vanishes (see Fig. W3). Compare the result with that of Section W4.

11. Consider a problem of constrained minimum pressure drag, when the volume of the body of revolution is prescribed. Use the method of Lagrange multipliers and the approach given in Section D26.

12. Consider the set of functions w, for which the integral

$$I = \int_V w^2 \, dv$$

exists. Is this set linear?

13. Prove that the product of symmetric permutable operators is a symmetric operator.

14. Check whether the functions $\sin(n\pi x/a)\sin(m\pi y/b)$, $n, m = 1, 2, \ldots$, are orthogonal over the domain $0 \le x \le a$, $0 \le y \le b$.

Chapter F

1. Find the values taken on by the interpolation functions of the beam element at the nodes. (Use `vfem33`.)

2. Find the values taken on by the interpolation functions of the bar element at the nodes. (Use `vfem33`.)

3. In view of the equation (F1.10), interpret the solutions to Problems 1 and 2.

4. Find the nodal forces due to the displacement $\{w\} = [0, a, 0, a]^T$ of the bar element. Interpret the answer. (Use `vfem33`.)

5. Check whether the mass matrix of the bar element depends on its orientation. (Use `vfem34`.)

6. Analyse the clamped beam represented by a single element. Interpret the answer.

7. Derive the stiffness and mass matrices of the composite beam, which consists of the three elements. (Use `vfem37`.)

8. Analyse a cantilever by the technique of reduced quantities. (Use `vfem37`.)

9. Analyse free vibrations of the clamped composite beam using three finite elements and recover numerically the results for the uniform beam. (Use vfem37.)

10. Analyse a cantilever plate using two plate elements. (Use vfem40.)

11. Find the values taken on by the interpolation functions of the plate element at the nodes. (Use vfem40.)

12. Solve the steady-state and transient problems of heat transfer in the fin (Sections F26 and F28), when the base temperature is $T_b = 60°C$. (Use vfem43.)

13. Instead of the two-element bar shown in Fig. F35 (Section F36), consider a similar three-element bar and obtain the solution with the help of (a) Lagrange multipliers, (b) penalty parameter.

14. What is the physical meaning of the Lagrange multiplier in the case of the problem shown in Fig. F35 (Section F36)?

Chapter S

1. Solve the first-order equation

$$Y_{,x} = x + Y \qquad 0 \leq x \leq 1$$

with the initial condition $Y(x = 0) = 0$ by the explicit and implicit Euler methods. Use a various step size, $X = 1, 0.25, 0.05, 0.01, 0.0025, 0.001, 0.00025, 0.0001$. Analyse the results.

2. Solve the above problem by the Runge–Kutta method of the third and fourth orders and analyze the results.

3. Integrate the equation

$$Y_{,x} - 3x^2 - x + 4 = 0$$

with the initial condition $Y(x = 0) = 0$ using the third-order Runge–Kutta scheme with $X = 1/2$ and $X = 1/4$.

4. Consider a linear undamped one-degree system subjected to a unit step force at $t = 0$ and obtain the solution by the fourth order Runge–Kutta method (specify the coefficients of the equation as unity). Investigate the convergence by using various step sizes.

5. Consider a linear undamped one-degree system subjected to a unit step force at $t = 0$ and obtain the solution by the multi-step methods given in Sections S7 and S8 (specify the coefficients of the equation as unity). Investigate the convergence.

6. Consider a linear undamped one-degree system subjected to a unit step force at $t = 0$ and obtain the solution by the predictor–corrector method (specify the coefficients of the equation as unity). Investigate the convergence.

7. Consider a non-linear pendulum treated in Section S11 but with the initial conditions given by $Y(x = 0) = 0, Y_{,x}(x = 0) = 2.618$ rad/s.

8. Solve the equation

$$Y_{,xx} + 2Y_{,x} = \cos x/(1 + 4\sin 2x)^{1/2}$$

on the interval [1, 2] with $Y(x = 1) = 2$, $Y_{,x}(x = 1) = 0$ by one of Newmark's methods, using the step-size $X = 0.25, 0.05, 0.01, 0.001$.

APPENDIX A: Manipulations with Matrices

Below the square brackets denote a rectangular matrix or a row and curled brackets a column. Two matrices $[a] = [a_{ij}]$ and $[b] = [b_{ij}]$ are equal if for all occurring i and j

$$a_{ij} = b_{ij}$$

If the matrices have the same number of rows and the same number of columns, their sum $[c]$ is

$$c_{ij} = a_{ij} + b_{ij}$$

The product of an $(m \times n)$ matrix $[a_{ij}]$ by a scalar s is merely the matrix $[sa_{ij}]$. The transpose $[a_{ij}]^T$ is the $(n \times m)$ matrix $[a_{ji}]$. For a symmetric matrix with real elements $[a_{ij}] = [a_{ij}]^T$.

A square matrix whose elements below and above the diagonal are all zero is said to be a diagonal one, and, further, if the diagonal elements are all 1, the matrix is a unit matrix.

Let $[a_{kj}]$ be an $(m \times n)$ matrix and $[b_{jp}]$ an $(r \times z)$ matrix. Then their product $[c] = [a][b]$ exists if $r = n$ and is the $(m \times z)$ matrix given by

$$c_{kt} = \sum_{j=1}^{N} a_{kj} b_{jt}$$

In general, $[a][b] \neq [b][a]$ and $[a][b] = 0$ does not imply $[a] = 0$ or $[b] = 0$. Some useful relations, often used in the text, are

$$([a][b])^T = [b]^T [a]^T$$

and

$$\{a\}^T \{b\} = \{b\}^T \{a\}$$

Note that the last equation specifies a scalar.

Consider the quadratic form

$$F = \{T\}^T [k] \{T\}/2$$

with a symmetric $[k]$. Then, if F is differentiated with respect to each of T_j, one gets

$$\left[\frac{\partial F}{\partial T_1} \frac{\partial F}{\partial T_2} \cdots \frac{\partial F}{\partial T_n}\right]^T = [k]\{T\}$$

Appendix B: Section-file Correspondence

Section	Maple file vfem*.ms	Mathematica file vfem*.ma
M1	1	–
M2	1	–
M3	2	–
M4	3	–
M5	4	–
M6	–	1
M7	–	2
M8	–	3
M9	–	4
D5	18	18
D6	19	19
D7	20	20
D8	20	20
D16	25, 25c	25
D18	9	9
D19	10	10
D20	12	12
D21	13, 14	13
D26	15	–
D27	32	32
F1	33	33
F2	33	33
F3	33	33
F4	33	33
F5	33	33
F6	34	34
F9	35	35
F10	35	35
F11	35	35

278 APPENDIX B

Section	Maple file vfem*.ms	Mathematica file vfem*.ma
F12	36	36
F13	36	36
F14	37	37
F15	37	–
F16	38	38
F17	39	39
F19	40	40
F20	41	41
F21	42	42
F22	42	42
F25	43	43
F26	43	43
F27	43	43
F28	43	43
F29	44	44
F31	44	44
F32	45	45
F34	46	46
F35	45	45
F36	47	47
F37	47	47
F40	48, 49, 50, 51	48, 49, 50, 51
S3	53	53
S5	54	54
S6	55	55
S8	56, 57, 58	56, 57, 58
S9	58	58
S11	62, 63	62, 63
S12	64	64
S15	65	65
S24	66	66
S25	67	67
S27	68	68
S30	69	69
W1	22	22
W4	23	23
W5	24	24
W8	26	26
W9	27	27
W10	28	28
W11	29	29
W12	30	30

Further Reading

The list of relevant sources is almost endless and only those intimately related to the subject are given below.

Chapter M

A. I. Beltzer (1990) Engineering analysis via symbolic computation – a breakthrough. *Appl. Mech. Revs.* **43** (6), (1991) 119–127.

B. W. Char, K. O. Geddes, G. H. Gonnet, B. L. Leong, M. B. Monagan and S. W. Watt (1991) *Maple V Language Reference Manual*. New York: Springer-Verlag.

B. W. Char, K. O. Geddes, G. H. Gonnet, B. L. Leong, M. B. Monagan and S. W. Watt (1991) *First Leaves: A Tutorial Introduction to MAPLE V*. New York: Springer-Verlag.

J. H. Davenport, Y. Siert and E. Tournier (1988) *Computer Algebra*. London: Academic Press.

S. Wolfram (1991) *Mathematica*. Redwood City, CA: Addison-Wesley.

Chapter D

A. I. Beltzer (1990) *Variational and Finite Element Methods – A Symbolic Computation Approach*. Berlin: Springer-Verlag.

M. L. Langhaar (1962) *Energy Methods in Applied Mechanics*. New York: Wiley.

S. G. Mikhlin (1964) *Variational Methods in Mathematical Physics*. Oxford: Pergamon.

K. Poltorak (1990) Cross-approximation method for solving dynamics problems of arbitrarily shaped plates. *J. Appl. Mech.*, **57**, 370–375.

R. S. Schechter (1967) *The Variational Method in Engineering*. New York: McGraw-Hill.

B. Vujanovic (1976) The practical use of Gauss' principle of least constraint. *J. Appl. Mech.* **3**, 491–497.

Chapter F

A. I. Beltzer (1990) *Variational and Finite Element Methods – a Symbolic Computation Approach*. Berlin: Springer-Verlag.
R. D. Cook, D. S. Malkus and M. E. Plesha (1989) *Concepts and Applications of Finite Element Analysis*. New York: Wiley.
H. Kardestuncer (editor) (1987) *Finite Element Handbook*. New York: McGraw-Hill.
R. H. MacNeal (1994) *Finite Elements: Their Design and Performance*, New York: Marcel Dekker.
J. N. Reddy (1985) *An Introduction to the Finite Element Method*. New York: McGraw-Hill.
F. L. Stasa (1985) *Applied Finite Element Analysis for Engineers*. New York: CBS Publ.
O. C. Zienkiewicz (1986) *The Finite Element Method*. London: McGraw-Hill.

Chapter S

N. S. Bakhvalov (1977) *Numerical Methods*. Moscow: Mir Publishers.
B. Carnahan, H. A. Luther and J. O. Wilkes (1969) *Applied Numerical Analysis*, New York: Wiley.
T. S. Parker and L. O. Chua (1989) *Practical Numerical Algorithms for Chaotic Systems*. New York: Springer-Verlag.
W. Shyy (1994) *Computational Modeling for Fluid Flow and Interfacial Transport*. Amsterdam: Elsevier.

Index

algorithm
  backward Euler, 194
  forward Euler, 194
array, 1, 9
assignment, 1, 15

boundary conditions, 37
  imposed, 32, 252
  natural, 32, 252

consistency, 230
constraints, 54, 182, 184
coordinate system, 95
  global, 95
  local, 95
  natural, 163
  serendipity, 168
characteristic equation, 218

database, 189
duct, 256
Duffings equation, 264

echo, 2
eigenvalue, 92
Euler–Lagrange equation, 28

fin, 152, 154
finite element, 77
  axial load, 171
  bar (truss), 85, 108
  beam, 77, 82, 114, 116, 117
  infinite, 181
  isoparametric, 170, 173, 175
  lineal, 163
  plane, 140, 145
  plate, 132
  rectangular, 145, 169, 188, 191
  singular, 179

  subparametric, 173
  superparametric, 173
  tetrahedron, 167
  triangular, 140, 165, 191
floating-point number, 1
foundation
  elastic, 35, 36
function
  admissible, 26
  Heaviside, 8
  trial, 33, 38, 48
functional, 26, 27
functions
  interpolation (shape), 80, 81

Jacobian, 174, 176, 178, 222

Lagrange multipliers, 182
Lagrangian, 29
locking, 189

matrix, 9, 19, 22
  assembly, 99, 101, 105, 117
  element mass, 81, 87, 96
  element stiffness, 77, 87, 96
  submatrix, 9
method
  Bubnov–Galerkin, 26, 33, 42, 150, 248, 262
  Kantorovich, 50
  Newmark, 215
  least squares, 267
  Raleigh–Ritz, 26, 36, 42, 251, 253, 258, 259
  Runge–Kutta, 199, 201
  predictor–corrector, 211, 213
  weighted residual, 34, 49
minimizing sequence, 48

operator, 44
  identity, 44
  inverse, 44
  linear, 44
  null, 44
  positive definite, 45
  symmetric, 45
optimization, 120

penalty parameter, 184
principle
  Gauss, 67, 266, 268
  Hamilton, 28

rotation-of-axes transformation, 96, 97

save, 2
scheme
  Adams–Bashforth, 206, 207
  Adams–Moulton, 207
  explicit, 197, 226, 234
  Gear, 207
  implicit, 197, 228, 235
  multistep, 205, 206, 209
  trapezoidal, 199
shaft, 241, 244
solve, 11, 21
stability, 31, 219, 229
stagnation flow, 268
start-up, 209

transversality condition, 64
truncation error, 195

variation, 28, 32

writeto, 2

WARNING

BY OPENING THIS SEALED PACKAGE YOU ARE AGREEING TO BECOME BOUND BY THE TERMS OF THIS LICENCE AGREEMENT which contains restrictions and limitations of liability.

IF YOU DO NOT AGREE with the terms set out here DO NOT OPEN THE PACKAGE OR BREAK THE SEAL. RETURN THE UNOPENED PACKAGE AND DISKS TO WHERE YOU BOUGHT THEM AND THE PRICE PAID WILL BE REFUNDED TO YOU.

THE MATERIAL CONTAINED IN THIS PACKAGE AND ON THE DISKS IS PROTECTED BY COPYRIGHT LAWS. HAVING PAID THE PURCHASE PRICE (WHICH INCLUDES A LICENCE FEE) AND IF YOU AGREE TO THE TERMS OF THIS LICENCE BY BREAKING THE SEALS ON THE DISKS, YOU ARE HEREBY LICENSED BY ACADEMIC PRESS LIMITED TO USE THE MATERIAL UPON AND SUBJECT TO THE FOLLOWING TERMS:

1. The licence is a non-exclusive right to use the written instructions, manuals and the material recorded on the disk (all of which is called the "software" in this Agreement). This licence is granted personally to you for use on a single computer (that is, a single Central Processing Unit).

2. This licence shall continue in force for as long as you abide by its terms and conditions. Any breach of the terms and conditions by you automatically revokes the licence without any need for notification by the copyright owner or its authorised agents.

3. As a licensee of the software you do not own what has been written nor do you own the computer program stored on the disks and it is an express condition of this licence that no right in relation to the software other than the licence set out in this Agreement is given, transferred or assigned to you.

   Ownership of the material contained in any copy of the software made by you (with authorisation or without it) remains the property of the copyright owner and unauthorised copies may also be seized in accord with the provisions of the copyright laws.

4. Copying of the software is expressly forbidden under all circumstances except one. The one exception is that you are permitted to make ONE COPY ONLY of the software on the disks for security purposes and you are encouraged to do so. This copy is referred to as the back-up copy.

5. You may transfer the software which is on the disks between computers from time to time but you agree not to use it on more than one computer at any one time and you will not transmit the contents of the disks electronically to any other. You also agree that you will not, under any circumstances, modify, adapt, translate, reverse engineer, or decompile the software or attempt any of those things.

6. Neither the licence to use the Software nor any back-up copy may be transferred to any other person without the prior written consent of Academic Press Limited which consent shall not be withheld provided that the recipient agrees to be bound by the terms of a licence identical to this licence. Likewise, you may not transfer assign, lend, rent, lease, sell or otherwise dispose of the software in any packaging or upon any carrying medium other than as supplied to you by Academic Press Limited. The back-up copy may not be transferred and must be destroyed if you part with possession of the software. Failure to comply with this clause will constitute a breach of this licence and may also be an infringement of copyright law.

7. It is an express condition of the licence granted to you that you disclaim any and all rights to actions or claims against Academic Press Limited and/or the copyright owner with regard to any loss, injury or damage (whether to person or property and whether consequential, economic or otherwise) where any such loss, injury or damage arises directly or indirectly as a consequence (whether in whole or part):

   a) of the failure of any software contained in these disks to be fit for the purpose it has been sold whether:
      i) as a result of a defect in the software, howsoever caused; or
      ii) any defect in the carrying medium howsoever caused; and/or

   b) any other act or omission of the copyright owner or its authorised agents whatsoever.

   and you agree to indemnify Academic Press Limited and the copyright owner against any and all claims, actions, suits or proceedings brought by any third party in relation to any of the above matters.

8. This agreement is to be read as subject to all local laws which are expressed to prevail over any agreement between parties to the contrary, but only to the extent that any provision in this Agreement is inconsistent with local law and no further.

9. This agreement shall be governed by the laws of the United Kingdom.